COMBAT
STUDIES
INSTITUTE

Research Survey No. 6

AD-A211 838

I0225151

A Historical Perspective on Light Infantry

by Major Scott R. McMichael

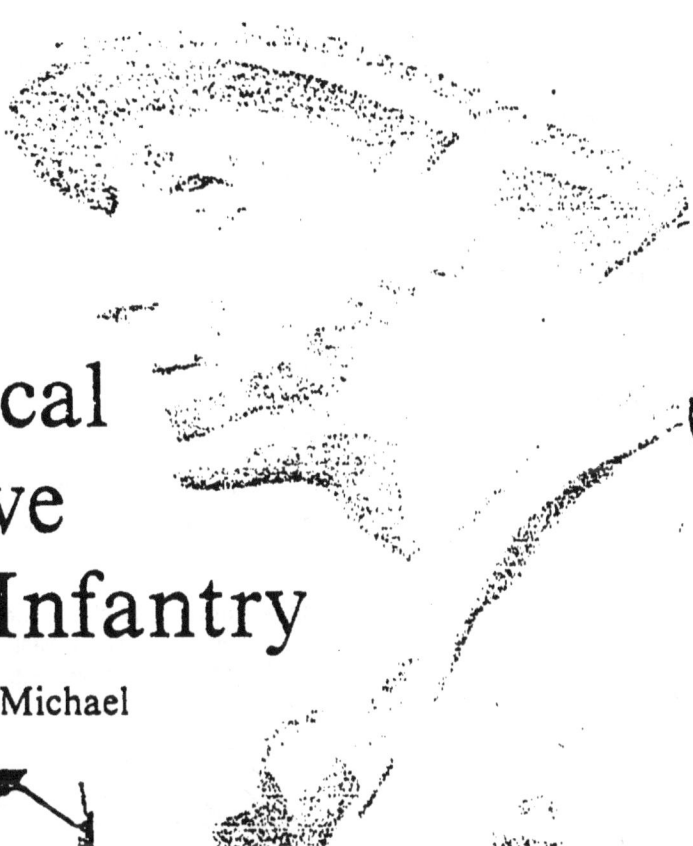

is document contains
ank pages that were
t filmed.

89 8 24 112

FOREWORD

The U.S. Army's commitment to light divisions is testimony to the importance of light infantry in modern war. The continuing usefulness of light forces goes beyond their ease of deployment. Light infantry exemplifies a state of mind that reveals itself in a unique tactical style, versatility, and élan that are so vital in battle. While the structure of light infantry makes it admirably equipped to fight in restricted terrain, it operates at considerable disadvantage in areas more suited to heavy forces. As with any military organization, commanders must consider both the capabilities and limitations of light infantry before committing it to battle.

Major Scott R. McMichael provides a valuable historical perspective for understanding the characteristics, organization, and operations of light infantry forces. Major McMichael's *Research Survey* examines four light infantry forces operating in varying settings: the Chindits in the 1944 Burma campaign against the Japanese; the Chinese Communist Forces during the Korean War; British operations in Malaya and Borneo from 1948 to 1966; and the First Special Service Force in its battles in the mountains of Italy during World War II. These examples are diverse in terms of time, areas of operations, and opposing forces, yet they reveal common characteristics of light forces and their operations.

A Historical Perspective on Light Infantry is based on extensive research in primary and secondary historical sources. The author has uncovered numerous doctrinal and operational manuals and reports and has gone beyond them to explore the more personal side of light infantry operations. This study is both fascinating reading and a valuable historical analysis of the capabilities and limitations of light infantry when faced with the test of battle.

September 1987

GORDON R. SULLIVAN
Major General, U.S. Army
Deputy Commandant

CSI *Research Surveys* are doctrinal research manuscripts, thematic in nature, that investigate the evolution of specific doctrinal areas of interest to the U.S. Army. *Research Surveys* are based on primary and secondary sources and provide the foundation for further study of a given subject. The views expressed in this publication are those of the author and not necessarily those of the Department of Defense or any element thereof.

Research Surveys ISSN 0887-235X

Research Survey No. 6

A Historical Perspective on Light Infantry

by Major Scott R. McMichael

U.S. Army
Command and General
Staff College
Fort Leavenworth, KS 66027-6900

Library of Congress Cataloging-in-Publication Data

McMichael, Scott R. (Scott Ray), 1951—
    A historical perspective on light infantry.

    (Research survey / Combat Studies Institute ; no. 6)
    Includes bibliographies.
    Supt. of Docs. no.: D 114.2:In 3
    1. Infantry. 2. Infantry drill and tactics.
3. Military history, Modern—20th century. I. U.S. Army
Command and General Staff College. Combat Studies
Institute. II. Title. III. Title: Light infantry.
IV. Series: Research survey (U.S. Army Command and
General Staff College. Combat Studies Institute) ;
no. 6.

UD145.M37   1987   356'.1   87-600239

For sale by the Superintendent of Documents, U.S. Government Printing Office, Washington, D.C. 20402

CONTENTS

Chapter 3. British Operations in Malaya and Borneo, 1948—1966

Part I. The Malayan Emergency

Part II. The Confrontation With Indonesia

Chapter 4. The First Special Service Force

ILLUSTRATIONS

Figures

Maps

TABLES

INTRODUCTION

What is the precise meaning of the term "light infantry"? How does light infantry differ from regular or conventional infantry? Are light infantry and dismounted infantry synonymous? Is light infantry merely conventional infantry given a light organization by stripping out heavy equipment and vehicles, or is it something quite different in terms of tactical style, attitudes, and utility? Are light infantry forces specialized elite forces or not? Do light forces have utility in low-, mid-, and high-intensity conflict?[1] These questions and others have occupied the attention of planners and trainers in the U.S. Army since 1983 when the Chief of Staff of the Army decided to introduce light infantry divisions into the force structure. Four years later, most of these questions remain unanswered. However, two main bodies of opinion have formed.

On one side of the issue, participants in the debate advance the idea that the primary determinant of light infantry is its organization. Light infantry forces, they argue, are light because they possess no organic, heavy equipment. They fight on foot, in close terrain, employing tactics that do not vary significantly from tactics employed by conventional infantry (i.e., motorized and mechanized infantry) forced to dismount. The value of light infantry, according to this line of argument, is its strategic mobility. It can be moved rapidly to "hot spots" anywhere in the world. Its activities and capabilities once deployed are less important than its ability to deploy to respond immediately to a crisis. This body of opinion is reflected most vividly in Field Circular 71-101, *Light Infantry Division Operations*, which describes the light infantry division essentially as a general purpose force. In fact, large portions of the text of this circular are identical to the text contained in FM 71-100, *Armored and Mechanized Division Operations*.

In contrast to this view, another interpretation exists, mostly European in its context and origins, that distinguishes light infantry from conventional infantry primarily on the basis of attitude and tactical style. Light infantry, from this perspective, has been a continuous component of European military formations for almost 300 years. Originally appearing in the form of French *chasseurs*, Prussian *Jaegers*, and Austrian *Grenz* regiments, these European light forces were used initially in skirmishing, hit-and-run raids, ambushes, ruses, and as guards for the main forces. In contrast to the strict, drill-style maneuvers of the heavy infantry, these light infantrymen were fleet, nimble, and resourceful—capable of operating independently from the army. The development of light infantry in Europe was paralleled in the New World by the rise of similar light units, such as the 60th Regiment of Foot and the American Ranger companies, units raised for scouting, skirmishing, and countering the activities of the French and Indian irregulars.

The European concept of light infantry expanded during the wars of the Napoleonic era. From 1790 to 1815, light units proliferated, evolving to include light artillery and light cavalry, and assuming a wider role on the battlefield. Covering withdrawals, screening advances, confusing the enemy and keeping him off-balance, light units made their presence felt at Ulm, Jena, Auerstedt, and throughout Wellington's entire Peninsular campaign in Spain. Employment of light infantry by European powers has continued unabated into the present day.[2]

As a result of these long years of experience, the European viewpoint on light infantry holds that light infantry is, first of all, a state of mind, and secondarily, a product of organization. The light infantry leader's mind-set, or ethic, differs significantly from the mind-set of conventional infantry leaders according to this view. This distinct light infantry mind-set produces a unique tactical style not normally exhibited by conventional infantry.

The purpose of this study is to clarify the nature of light infantry more definitively. To support this goal, the general characteristics of light infantry forces will be identified, and an analysis of how light forces operate tactically and how they are supported will be presented. In the process, the relationship of the light infantry ethic to its organization will be evaluated, and the differences between light infantry and conventional infantry will be illuminated. For the purpose of this study, the term "conventional infantry" will refer to modern-day motorized and mechanized infantry and to the large dismounted infantry forces typical of the standard infantry divisions of World War II, the Korean War, and the Vietnam War.

This study concludes that light infantry is, in fact, unique and distinct. A light infantry ethic exists and manifests itself in a distinctive tactical style, in a special attitude toward the environment, in a freedom from dependence on fixed lines of communication, and in a strong propensity for self-reliance.

No such thing as a standard light infantry force exists: light infantry comes in all shapes and sizes. It has been employed in various environments by a variety of national armies. Understanding the nature of light infantry thus requires a thorough examination of light infantry forces in their diverse forms.

Consequently, this study is based on a historical analysis of four separate light infantry forces that were employed during and since World War II. Each case study is different, the forces having been selected for their diversity of size and organization and for the purposes for which they were used. These forces also exhibited differences in the intensity of their conflicts, the nature of their threats, and in the terrain and climate where they were employed.

Chapter 1 concerns the Chindits, a seven-brigade force commanded by Major General Orde Wingate in the 1944 Burma campaign against the Japanese. Composed of British, Gurkha, African, and American troops, the Chindits conducted large-scale, guerrilla-style interdiction against Japanese lines of communication in the jungles and mountains of northern Burma for a period of five months.

Chapter 2 addresses the operations of the Chinese Communist Forces (CCF) during the Korean War. The CCF was a light infantry army. Lacking the industry necessary to equip and transport its legions, the Chinese relied on a philosophy of "man over weapons" (i.e., manpower used to counter weapons superiority). Its enemy, the U.S. Army, in stark contrast, wielded the most technologically advanced army in the world. The mid-intensity war between the two powers, ranging over cold desolate mountains, barren hills, and frozen streams, resulted in a stalemate.

British operations in Malaya and Borneo from 1948 to 1966 are the subject of chapter 3. This case study explores the use of light infantry forces in protracted counterinsurgent warfare. Although the terrain and climate of Malaya and Borneo are similar to that of northern Burma, this case study differs significantly from the chapter on the Chindits in terms of the nature of the threat, the methods of organization, the level of intensity, and the tasks undertaken by the light forces.

The last case study moves out of Asia into the mountains of Italy. Chapter 4 examines the First Special Service Force (FSSF), an elite, Canadian-American, regimental-size light infantry force especially trained for amphibious assaults and operations in snow-covered mountains. The FSSF established a remarkable record of accomplishment in its two short years of existence from 1942—44, during which it was employed in a wide variety of roles.

Collectively, the four case studies represent a wide array of terrain and climate: jungle, swamp, tropical mountains, relatively dry mountains (Korea), cold areas, snow-covered mountains, and—in the case of the FSSF—isolated island strongholds. The threat ranges from strong German conventional divisions to small parties of Chinese terrorists. The sizes of the forces vary from army level to battalion level and smaller, with units differing in organization. The four case studies encompass elite and nonelite light forces from four different national armies—British, Canadian, Chinese, and American—that were involved in mid-intensity war, low-intensity conflict, and rear-area operations. Finally, the types of tactical operations discussed include amphibious assaults, reconnaissance and combat patrols, ambushes, deep raids, conventional attacks, defense of strongpoints, static linear defenses, area sweeps, interdiction, and economy of force operations.

These diverse conditions ensure that the study has the proper scope and comprehensiveness to permit the drawing of legitir... te conclusions on the nature of light infantry. Moreover, where appropriate, the study introduces the experience of other light infantry forces to reinforce or amplify specific points.

Each case study, while employing a fairly standard format, places the infantry force in a historical context, explaining how and why it came to be formed. The inquiry then examines each force's tactical environment, demonstrating its influence on light infantry operations. Next, the study discusses the selection, organization, and training of the force. Then, it evaluates the operational employment and tactical techniques of the force in the offense and defense. This is followed by an analysis of the force's combat support, leadership, and logistics. Each case study closes with a review of the problems experienced by light forces and presents a number of pertinent conclusions.

Chapter 5 synthesizes the information and conclusions presented in the four case studies. Identified in greater relief are the four components of the light infantry ethic. Also brought into greater focus are the principles that govern the organization, training, operational employment, tactics, combat support, logistics, leadership, and role of technology for light infantry forces. The chapter closes with a review of the problems and vulnerabilities of light infantry. The text concludes with a table comparing the distinctive differences between light infantry and conventional infantry.

This study does not directly address the issues of strategic deployability, contingency operations, and the heavy-light forces mix in Europe. It also generally steers away from evaluating specialized light infantry forces such as Rangers, Commandos, and airborne units. While these types of forces may properly be considered as light infantry, their costly, specialized training, privileged access to resources, and unusual capabilities place them on the fringe of the genre. This study does not ignore these forces: the FSSF is representative of the type, and a short discussion of the Special Air Service (SAS) is also found in chapter 3. The emphasis of the study, however, is on the less-specialized light infantry forces.

Finally, this study is not an argument for or against light infantry divisions. Even though it offers many insights into and implications for the training, organization, and employment of light forces today, only a few specific recommendations are made. In this regard, this study is descriptive, not prescriptive. On the other hand, this study provides a solid historical perspective on the essential nature of light infantry. It enumerates a number of principles concerning how successful light infantry forces have been organized, trained, and employed. Thus, it can serve as a repository of historically verified guidelines on the use of light infantry.

NOTES

Introduction

1. The terms "low-, mid-, and high-intensity conflict" are used periodically throughout this study. Their definitions are provided below. The definition of low-intensity conflict has been adopted officially by the U.S. Army. No official definitions of mid- and high-intensity conflict exist. The ones chosen below are commonly used as working definitions throughout the U.S. Army. They are from U.S. Department of the Army, FM 100-20, *Low Intensity Conflict* (Washington, DC, January 1981), 14; and FM 100-20, *Low Intensity Conflict*, (Fort Leavenworth, KS, 16 July 1986), v.

- Low-intensity conflict: a limited political-military struggle to achieve political, military, social, economic, or psychological objectives. It is often protracted and ranges from diplomatic, economic, and psychosocial pressures through terrorism and insurgency. It is generally confined to a geographic area and is often characterized by constraints on the weaponry, tactics, and levels of violence. Low-intensity conflict involves the actual or contemplated use of military capabilities up to, but not including, combat between regular forces.

- Mid-intensity conflict: a war between two or more nations and their respective allies, if any, in which the belligerents employ the most modern technology and extensive resources in intelligence; mobility; firepower (excluding nuclear, chemical, and biological weapons); command, control, and communications; and service support for limited objectives under definitive policy limitations as to the extent of destructive power that can be employed or the extent of geographic area that might be involved.

- High-intensity conflict: a war between two or more nations and their respective allies, if any, in which the belligerents employ the most modern technology and extensive resources in intelligence; mobility; firepower (including nuclear, chemical, and biological weapons); command, control, and communications; and service support.

2. This short summary of European light forces draws from David Gates, "Western Light Forces and Defence Planning. 1. Some Parallels from the Past," *Centrepiece* no. 8 (Aberdeen, Scotland: Centre for Defence Studies, Summer 1985), 1—32.

BIBLIOGRAPHY

Introduction

Gates, David. "Western Light Forces and Defence Planning. 1. Some Parallels from the Past." Centrepiece no. 8. Aberdeen, Scotland: Centre for Defence Studies, Summer 1985.

U.S. Department of the Army. FM 100-20. *Low Intensity Conflict*. Washington, DC, January 1981.

The Chindits

Introduction

The decisive campaign of the long land war against the Japanese in Burma in World War II was the Battle of Imphal-Kohima. For 100 days, from March through June 1944, the troops of the Fourteenth Army under Lieutenant General William Slim met their bitter Japanese enemies in a convulsive struggle for control of the eastern gates to India. Ultimately, this British multi-race army defeated the Japanese attack and began the slow task of clearing the invaders from northern and southern Burma. While this great battle was being contested, another war, smaller in scale but no less fierce, was being fought 200 miles in the Japanese rear. Here, over 20,000 specially trained jungle soldiers attempted to weaken the Japanese Army by delivering a knockout blow to its unprotected "guts." Three thousand of these troops were American volunteers, officially known as the 5307th Composite Unit (Provisional) and popularly known as "Merrill's Marauders" (although they referred to themselves as "Galahad"). The other larger part of this extraordinary collection of fighting men was the Chindits, also known as the Special Force.

Essentially, Galahad and the Chindits were light infantry jungle troops organized and trained for guerrilla-style interdiction against Japanese lines of communication. During the campaign in Burma, circumstance and misuse forced these units into the conventional roles of positional defense and direct assaults against strong enemy fortifications. Galahad and the Chindits also operated at the operational level of war in that their deployment into Burma and their tactical objectives contributed directly to the attainment of strategic goals. In fact, the Chindit War, as it is called by British military historian Brigadier Shelford Bidwell, is one of the best examples in recent history of light infantry forces employed at the operational level of war.

The record of the Chindit War is one of high drama, involving both exhilarating triumph and bewildering tragedy. Galahad, for example, gave Lieutenant General Joseph Stilwell his most notable success—the capture of the Myitkyina airfield; and yet his forces were ruined and destroyed in the process. Each of the five British Chindit brigades deployed in Burma suffered casualties of from 50 to 95 percent of its original force. In one case, the 111th Infantry Brigade stumbled out of its last engagement with only 118 men of its original 3,000 fit for further service.

A close study of Galahad and the Chindits is of high value because their operations form classic examples of light infantry tactics in close jungle terrain, deep in the enemy's rear. Moreover, a study of these forces demonstrates that light forces can be used at the operational level of war. Chindit operations also show how guerrilla and conventional tactics can be effectively mixed to achieve significant tactical objectives. Finally, the history of the Chindit War reveals the costs and dangers of misusing light forces in roles or situations for which they are not suited.

This study begins with a background section providing an overview of the land theater, including a description of the strategies and dispositions of the opposing forces in the spring of 1944. The nature of the participation of the Chindits and Galahad in the Burma campaign of 1944 is depicted. Then, the study describes the organization and training of the Chindits, their basic operational concept, Chindit tactics in their several variants, Chindit logistics, leadership and morale, and costs and problems. The study concludes with a brief assessment of the results achieved by the Chindits and an analysis of the lessons of their operations that remain relevant today.

Dispositions and Strategy

The dispositions of the antagonists in Burma in March 1944 are shown on map 1. In the north, Stilwell's U.S.-trained First Chinese Army of the 22d and 38th Divisions was opposed by Lieutenant General Tanaka's 18th (Chrysanthemum) Division, a crack outfit with extensive jungle experience and many battle honors. East of Myitkyina (pronounced Mitchinah), Chiang Kai-shek's Chinese divisions, never very aggressive, were held in check by the Japanese 56th Division. On either side of the mighty Chindwin River, the mixed British-Indian IV Corps of the Fourteenth Army defended against Lieutenant General Mutaguchi's Fifteenth Army, composed of three large infantry divisions. Farther south, in the province of Arakan, the British XV Corps faced the Japanese 55th Division.

In the far north, an Allied headquarters at Fort Hertz coordinated the activities of a fair-size force of Kachin levies—a jungle guerrilla force operating in a decentralized mode under Office of Strategic Services (OSS) leadership against the rear elements and garrisons of the 18th Division.[1] This huge land theater encompassed some 35,000 square miles of some of the most difficult terrain for warfare in the world.

The Quebec Conference of August 1943 between the leaders of the United States and Great Britain established the Allied strategy for Burma in 1944. The strategy envisioned the reconquest of northern Burma by Chinese and American forces under Stilwell's Northern Combat Area Command (NCAC). Stilwell hoped to force the Japanese 18th Division out of its positions in the Hukawng and Mogaung valleys, ultimately pushing through to Myitkyina, drawing the Ledo road behind him. Stilwell also hoped that Chiang's divisions in Yunnan province east of Myitkyina would conduct their own offensive from the opposite direction. Once a land link to China was restored and the Ledo-Yunnan road completed, China would then become the main theater for the final defeat of the Japanese Army. More or less simultaneously with Stilwell's advance in North Burma, General Slim's Fourteenth Army intended to initiate its own major offensive from the central front in Assam against the bulk of the Japanese Army in Burma.

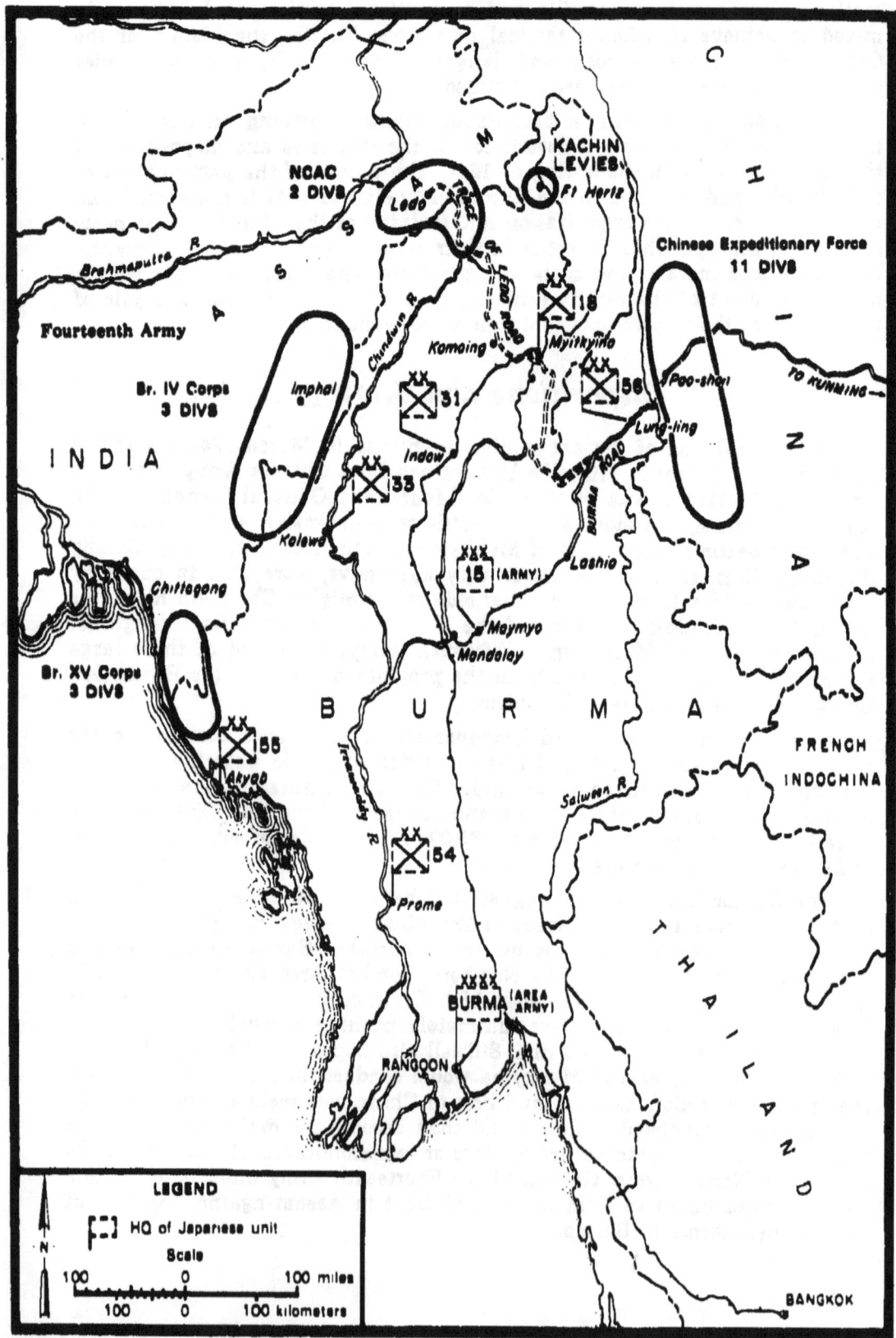

NCAC
2 DIVS

KACHIN
LEVIES

Ft Hertz

Ledo

Chinese Expeditionary Force
11 DIVS

Fourteenth Army

Brahmaputra R.

Chindwin R.

Kamaing

Myitkyina

18

Br. IV Corps
3 DIVS

Imphal

31

56

Pao-shan

TO KUNMING

INDIA

Indaw

33

Kalewa

Lung-ling

BURMA ROAD

15 (ARMY)

Lashio

C H I N A

Chittagong

Maymyo
Mandelay

Salween R.

FRENCH
INDOCHINA

Br. XV Corps
3 DIVS

55

Akyab

B U R M A

Irrawaddy R.

54

Prome

THAILAND

BURMA (AREA ARMY)

RANGOON

BANGKOK

LEGEND

N

HQ of Japanese unit

Scale

100   0   100 miles

100   0   100 kilometers

Source: Romanus and Sunderland, *Stilwell's Command Problems* 120

Map 1. Disposition of forces, Burma, March 1944

In addition to developing the strategy described above, the Quebec Conference authorized the creation of one American and six British brigades to be employed in Burma as long-range penetration groups (LRPG). Under the command of British Major General Orde C. Wingate, these groups were to interdict the Japanese rear in order to support the three contemplated Allied offensives. In 1943, Wingate had already organized, trained, and led one LRPG, the 77th Infantry Brigade, in extended operations in Japanese-held territory.

This force had infiltrated deep into the enemy rear on foot and, over the course of several weeks, blew railway bridges, mounted ambushes, destroyed supply dumps, and conducted other similarly disruptive activities—slipping into the safety of the jungle after each attack. But forced by a determined Japanese response to exfiltrate to their own lines, the 77th Brigade suffered debilitating casualties. Fully one-third of the 3,000-member force was lost outright, while the majority of the survivors were found unfit for further service due to disease, injury, exhaustion, or malnutrition.[2]

While it must be acknowledged that the military value of this expedition was quite low, especially considering the terrible casualties incurred, Chindit I (the first British operation) provided an undeniable psychological boost to the British and Indian formations that had been driven so rudely and rapidly out of Burma in 1942.[3] The expedition restored their confidence. Wingate had proven that the Japanese were not invincible, that they could be defeated at their own game—jungle fighting. Moreover, he had demonstrated that a sizable force could be maintained behind enemy lines and could be effective under the right circumstances. The major flaw in the expedition was that the operations of the 77th Brigade were not conducted in concert with a simultaneous attack by the main armies.

Chindit II (the second major British operation) was to be somewhat of a repetition of Chindit I, with four notable differences. First, Chindit II obviously was being mounted on a much larger scale, magnified by a factor of seven to one. Second, unlike Chindit I, Wingate's long-range penetration in 1944 was to be coordinated to complement and contribute directly to the advance of the main Allied armies by cutting Japanese lines of communication in the three directions described earlier. Third, in 1944, Wingate's brigades were to be supported by a dedicated air force, the No. 1 Air Commando. Finally, due to the combination of factors above, the tactics employed by the Chindits in 1944 were to be altered, producing a blend of guerrilla and conventional tactics. These last innovations, however, were not always understood or appreciated by Wingate's subordinate commanders.

Basic Plan of Operation

On 4 February 1944, General Slim issued a directive to Wingate ordering the Special Force (i.e., the Chindits) to march and fly into the Indaw-Railway valley area to accomplish the following missions:

> (1) Help the advance of Stilwell's force on Myitkyina by cutting the communications of the Japanese 18th Division, harassing its rear and preventing its reinforcement.
>
> (2) Create a favorable situation for the Yunnan Chinese to cross the Salween River and enter Burma.
>
> (3) Inflict the greatest possible damage and confusion on the enemy in North Burma.[4]

4

It should be noted that these missions differed slightly from the strategy laid down in 1943 in that they did not include a specific directive to cut the lines of communication to the Japanese forces on the central front in the west. The thrust of Wingate's mission was to support Stilwell's advance down the Hukawng and Mogaung valleys and thence to Myitkyina.

Wingate's plan to fulfill Slim's directive was called Operation Thursday (see map 2). It provided for the secret insertion of three brigades deep into the enemy's rear—one by foot and two by air. The 16th Brigade, under Brigadier Fergusson, marched into Burma from just south of Ledo. Crossing the Chindwin River by glider-delivered assault boats, the brigade trekked 450 miles to a site known as Aberdeen, 27 miles northwest of Indaw, where it established a stronghold or permanent, defended base. From there, the 16th was to proceed to capture the Indaw airfields and destroy the defending Japanese garrison.

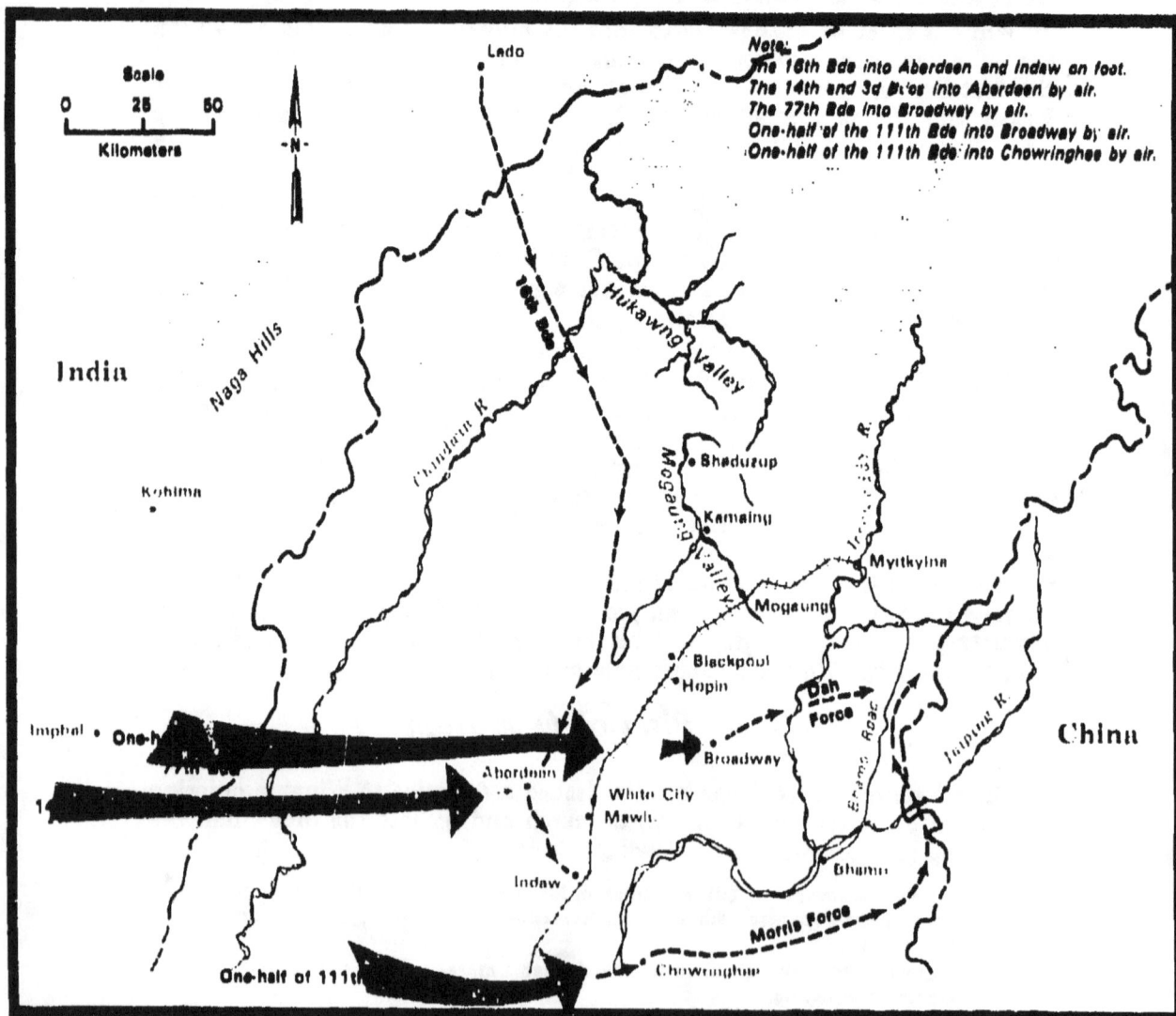

Map 2. Operation Thursday

5

As part of the operation, the 77th and the 111th Brigades flew by glider and C-47 to two landing zones—Broadway and Chowringhee—located east and northeast of Indaw. The 77th split its forces at that point, leaving one attached battalion to garrison its landing zone and stronghold (Broadway) and sending another called the Morris Force to interdict the Bhamo-Myitkyina road and Irrawaddy River—the main supply routes to the 56th Division. The main force of the brigade marched west and established a block on the Mandalay-Myitkyina railroad at White City. However, the 111th was initially delayed in its flight to the Chowringhee landing zone. It crossed the Irrawaddy River, also by glider-delivered assault boats, before moving west to harass Japanese rear elements in the area west of Indaw. The 111th also sent one column of one battalion to join the Morris Force.[5] Later in the campaign, the 111th established a block at Hopin on the Mandalay-Myitkyina railroad.

Meanwhile, Wingate held the 14th, 3d, and 23d Brigades in reserve to reinforce success and to relieve the first three deployed brigades when they became exhausted, needed assistance, or lost their effectiveness. Technically, Operation Thursday covered only the deployment and the initial objectives of the Special Force. Wingate retained authority to modify the objectives of his brigades depending on how the situation developed. Wingate also intended to withdraw his Chindits after ninety days of rear-area operations and before the monsoon season arrived in June. Chindit I had proven that operations of the type undertaken by the Chindits should not extend past ninety days. After that time, units in the enemy's rear lost their effectiveness.[6]

Major General Orde C. Wingate and his successor Brigadier W. D. A. "Joe" Lentaigne

Operation Thursday had been underway only a short time when several important events took place. First, the Allied offensive on the central front was delayed by an attack from the Japanese (see map 3). As a result, Chindit operations were not enhanced as expected by a simultaneous advance of the Fourteenth Army. However, Stilwell's offensive in the north continued as planned. The second event was Wingate's death in a plane crash before the end of March. His successor, Brigadier Lentaigne, did not possess Wingate's

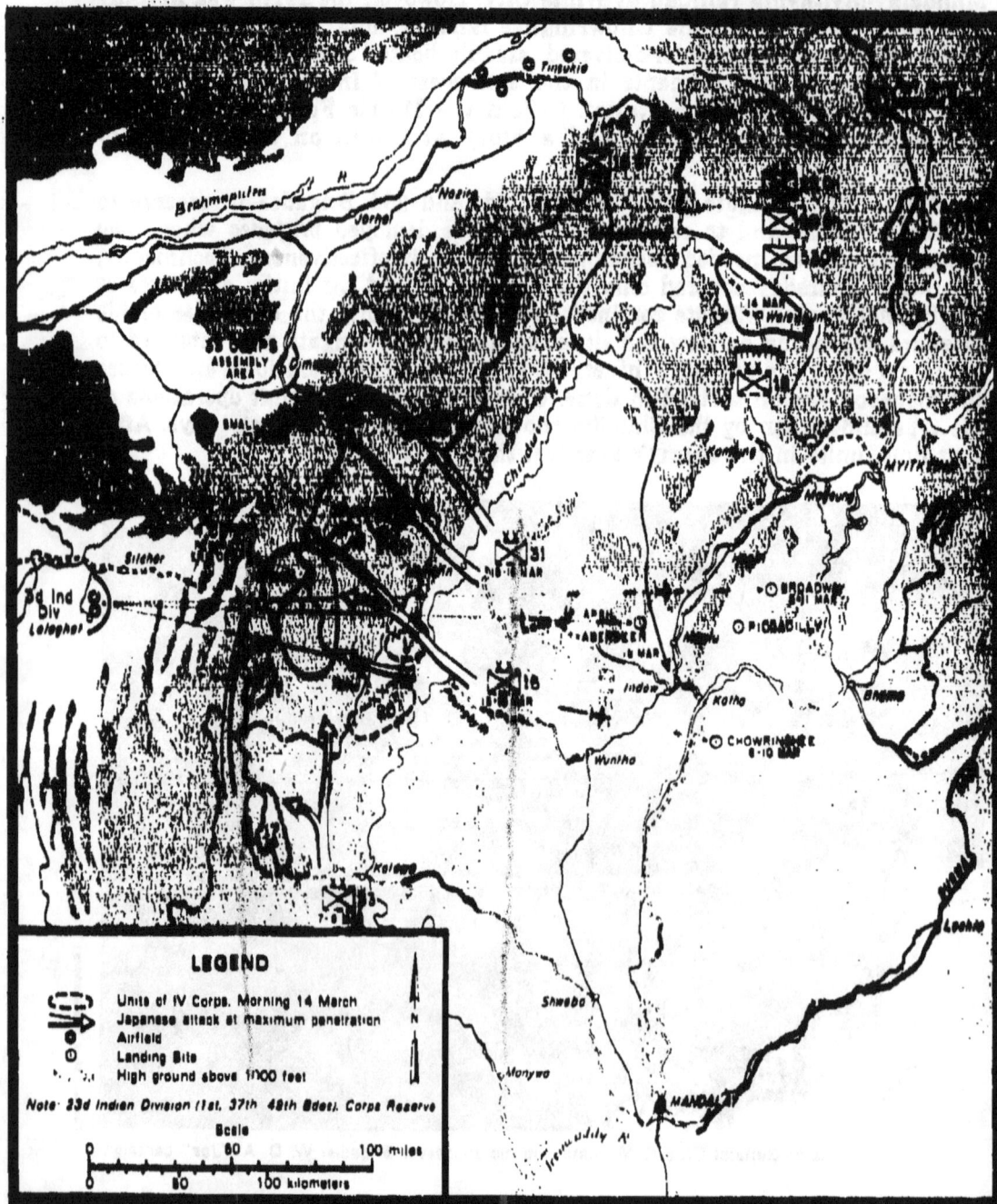

Source: Romanus and Sunderland, Stilwell's Command Problems, 173.

Map 3. Japanese offensive at Imphal, March—April 1944

vision or imagination. Wingate's death, coupled with the Japanese offensive, led to lessened efficiency in the use of the Chindit brigades. For instance, once it was clear that the decisive campaign in Burma was occurring at Imphal, rather than in the north, the Chindits should have been diverted in that direction, not to the north.

Operation Thursday did not include Galahad. Although initially intended to be under Wingate's command, Galahad passed to Stilwell, instead, on the basis of his adamant personal request to Lord Mountbatten, the theater commander.

Stilwell's plan for Galahad had a slightly different cast than Wingate's. As part of Stilwell's plan, Galahad's objectives were not to be as deep as those of the Special Force (at least initially) and were to be more closely coordinated with advances by Stilwell's First Chinese Army. In essence, Stilwell directed Galahad to conduct a series of deep envelopment operations, wherein the 5307th would march secretly around the right flank of the Japanese and establish blocks directly athwart the single main road in the enemy's rear, but close enough to the forward defenders to be a short-term threat. The plan called for the first blocks to be established at Walawbum in the Hukawng valley (see maps 4 and 5) and the second blocks to be near Shaduzup in the Mogaung valley (see map 6). The third mission of Galahad was to be a surprise attack on the airfield at Myitkyina (see map 7). In all three cases, once Galahad was in position, Stilwell intended to push his Chinese divisions hard against the Japanese, forcing them to divide their attention between his two forces. At the same time as Galahad and the First Chinese Army played this game of hammer and anvil, Stilwell and Slim hoped that the operations of the Special Force would severely degrade Tanaka's ability to sustain his hard pressed force. During the operation, Galahad was officially under the command of Brigadier General Frank K. Merrill, an old Stilwell hand. Actually, Colonel Charles N. Hunter, the deputy commander, usually directed Galahad in the field due to Merrill's poor health. Indeed, Merrill was evacuated three times during the campaign because of his weak heart.

The area where Galahad and the Chindits operated was a mosaic of rugged hills, saw-toothed ridges, high mountains, and noxious valleys, traversed by many small and large rivers bordered by thick tropical jungle. Few paths and trails existed, and maps often proved unreliable. The jungle included stands of bamboo so thick that a tunnel, instead of a path, had to be hacked out for the columns to pass through. Due to the numerous rivers, the Chindits made hundreds of river crossings, emerging from the water invariably speckled with leeches. During monsoon season, the area became almost impassable. Low ground became inundated and mountain sides so muddy and greasy that men had to crawl up over steps that were laboriously hacked out. The high humidity, constant rain, and high temperatures fostered heat prostration. Moreover, the mosquitos and mites infesting the area carried the germs of malaria and scrub typhus. Operating in this terrain required the highest levels of physical endurance and mental toughness, and every day spent on the march was torture.

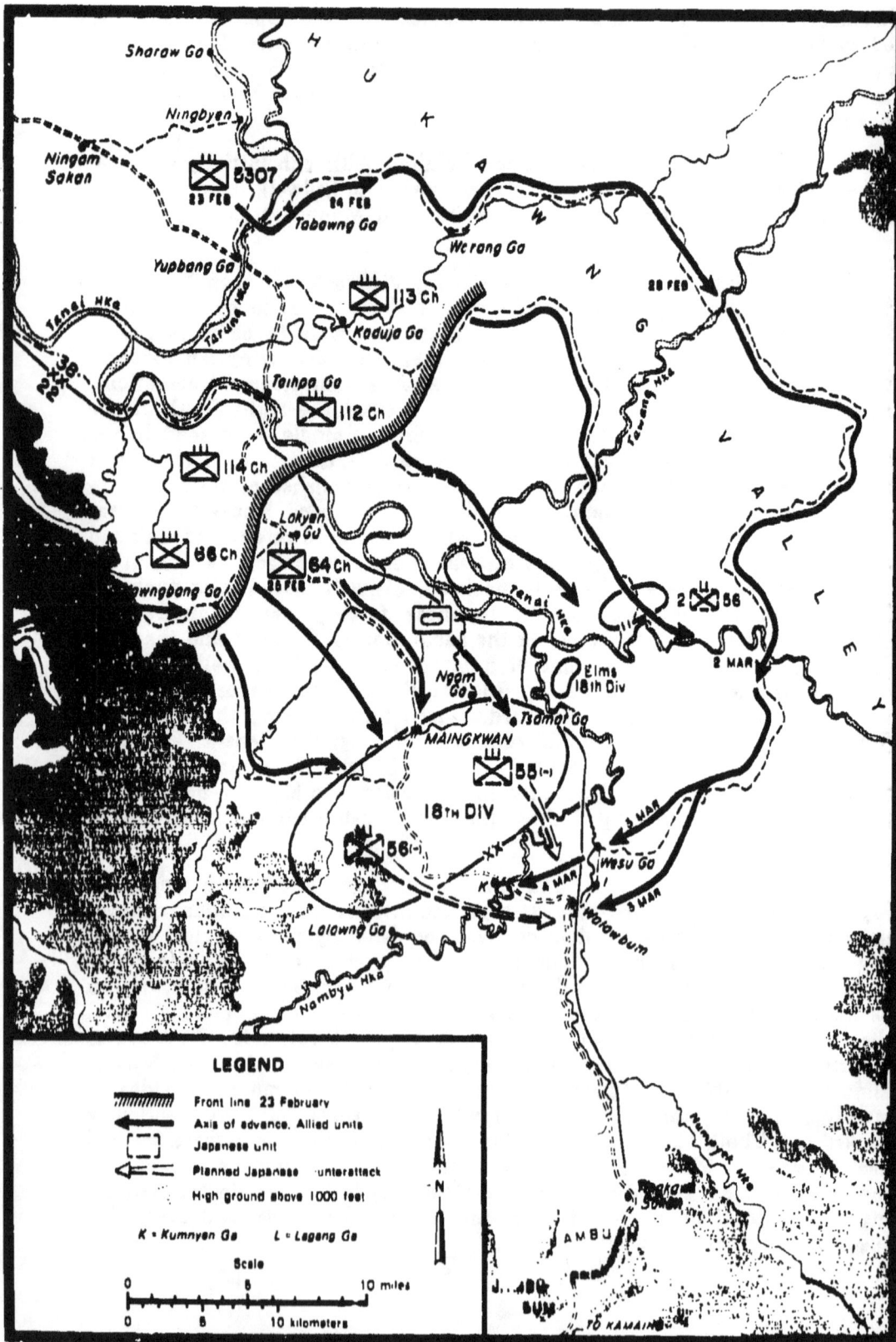

Source: Romanus and Sunderland, *Stilwell's Command Problems*, 144

Map 4. Galahad's advance to Walawbum, 23 February—4 March 1944

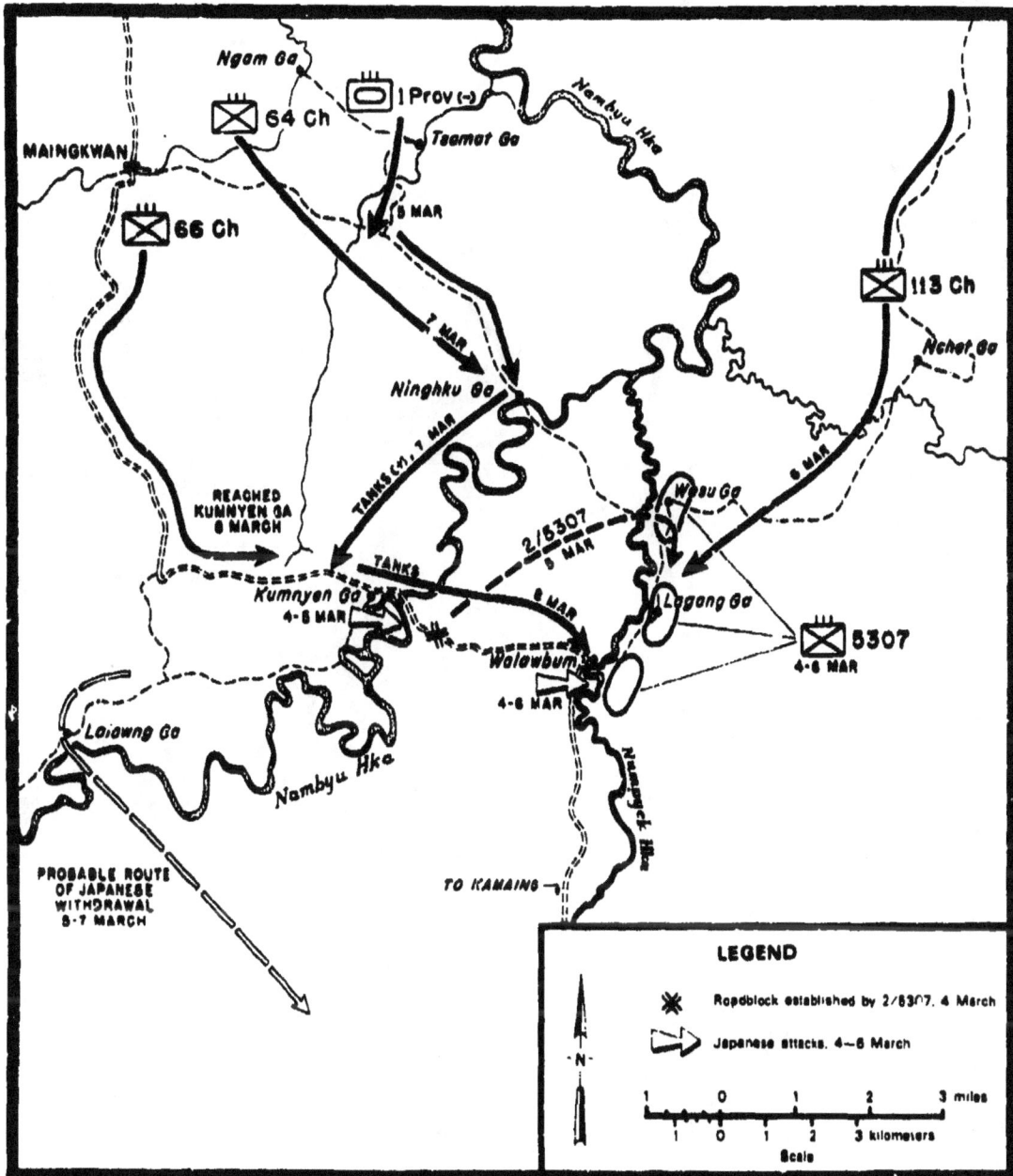

LEGEND

✳ Roadblock established by 2/5307, 4 March

⟾ Japanese attacks, 4—6 March

Source Romanus and Sunderland *Stilwell's Command Problems*, 151

Map 5 The fight at Walawbum, 4—8 March 1944

10

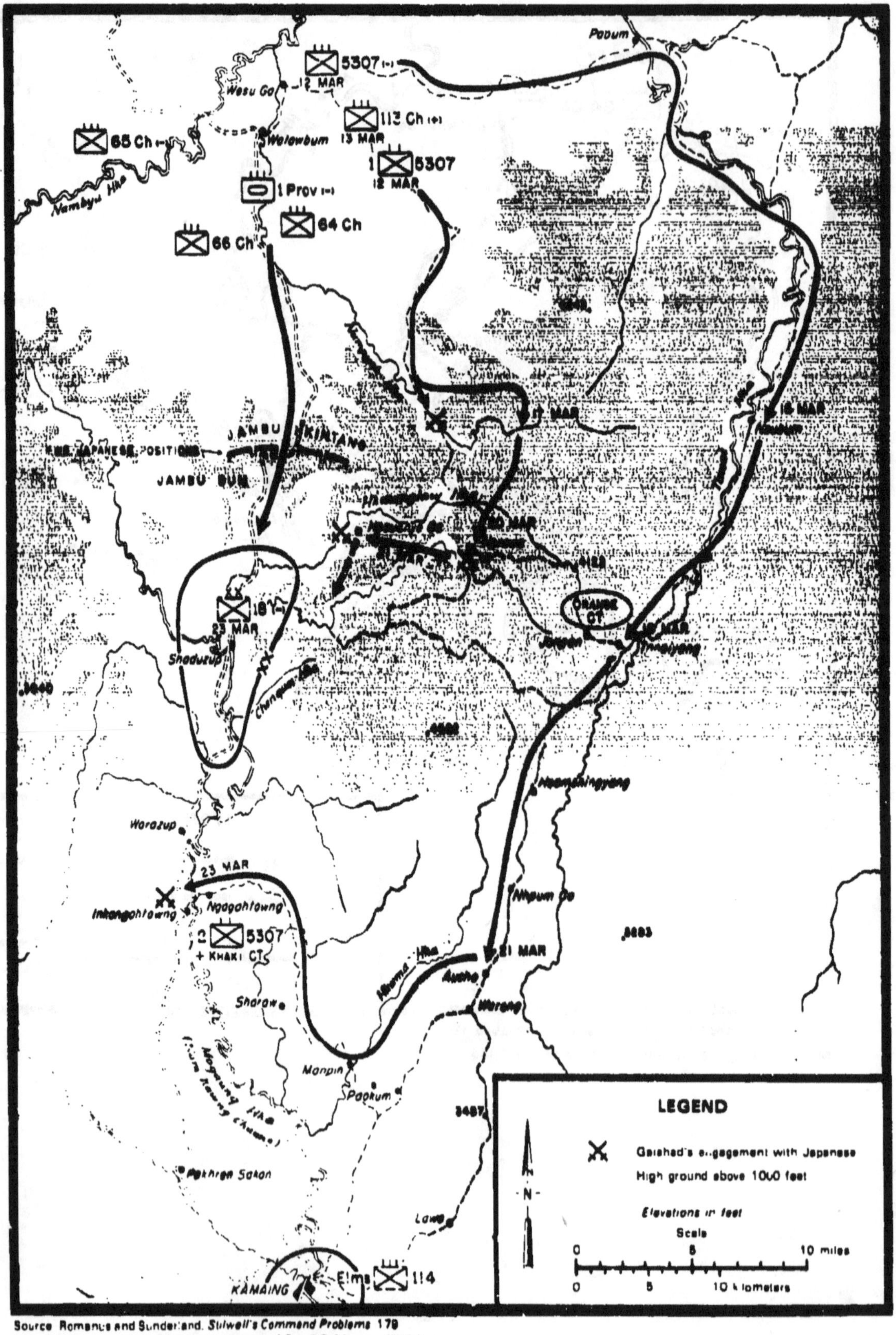

Source: Romanus and Sunderland. *Stilwell's Command Problems* 179

Map 6 Galahad at Inkangahtawng, 12—23 March 1944

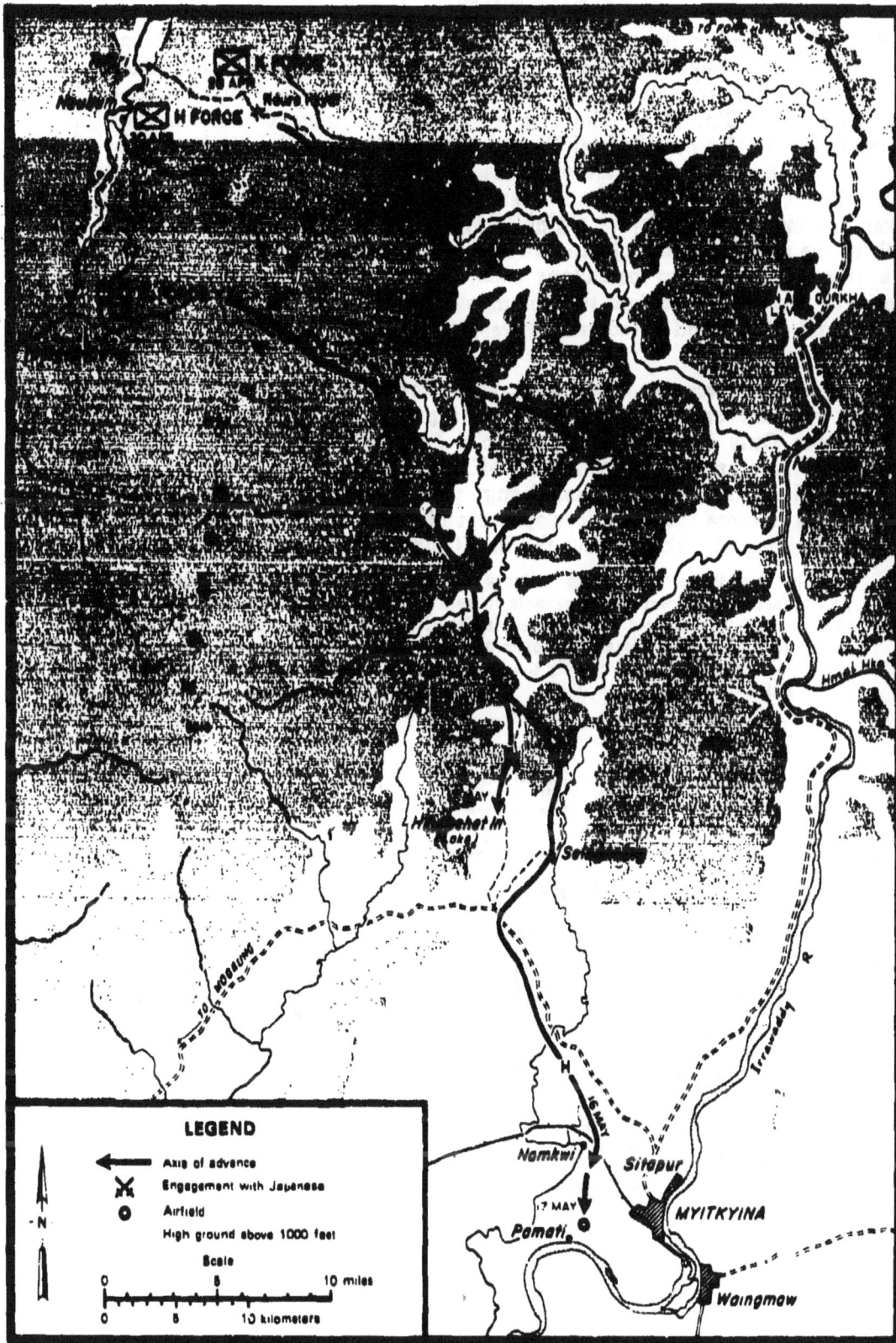

LEGEND

⟵ Axis of advance

Engagement with Japanese

○ Airfield

High ground above 1000 feet

Scale

0 ⎯⎯ 5 ⎯⎯ 10 miles

0 ⎯⎯ 5 ⎯⎯ 10 kilometers

Source Romanus and Sunderland. *Stilwell's Command Problems*. 224

Map 7 Galahad advances to Myitkyina, 28 April—17 May 1944

12

Organization

Wingate's Chindits, the men who were to enter this menacing environment, were known under several names, but the most commonly used title in 1944 was the Special Force. The 77th Infantry Brigade, employed in Chindit I, was retained for Chindit II. The 77th also provided many of the cadres for the expansion of the new Special Force into six brigades. The 111th and the 3d West African Brigades were organized separately as components of the Special Force, but the 70th Indian Infantry Division, a regular line unit, was broken up into three separate Chindit brigades—the 16th, 14th, and 23d— much to the distress of the old Indian Army bureaucracy. These six brigades were organized with four battalions each, as shown in figure 1. Each battalion was further divided into two columns commanded by the battalion commander and his second in command, respectively, for tactical operations in the jungle (see figure 1).

The Chindits were not elites; they were perfectly ordinary soldiers from perfectly ordinary battalions assigned to Wingate to be prepared for extraordinary tasks. Only 5 percent of this Special Force were volunteers.[7] Wingate, himself the most unorthodox of British officers, did not believe that a special kind of soldier was required for long-range penetration. Wingate believed that adept jungle fighters could be developed out of any unit through good leadership and training. Speaking about the first Chindit expedition, he declared:

> What was it that made these ordinary troops, born and bred for the most part to factories and workshops, capable of feats that would not have disgraced Commandos? The answer is that given imagination and individuality in sufficient quantities, the necessary minimum of training will always produce junior leaders and men capable of beating the unimaginative and stereotyped soldiers of the Axis.[8]

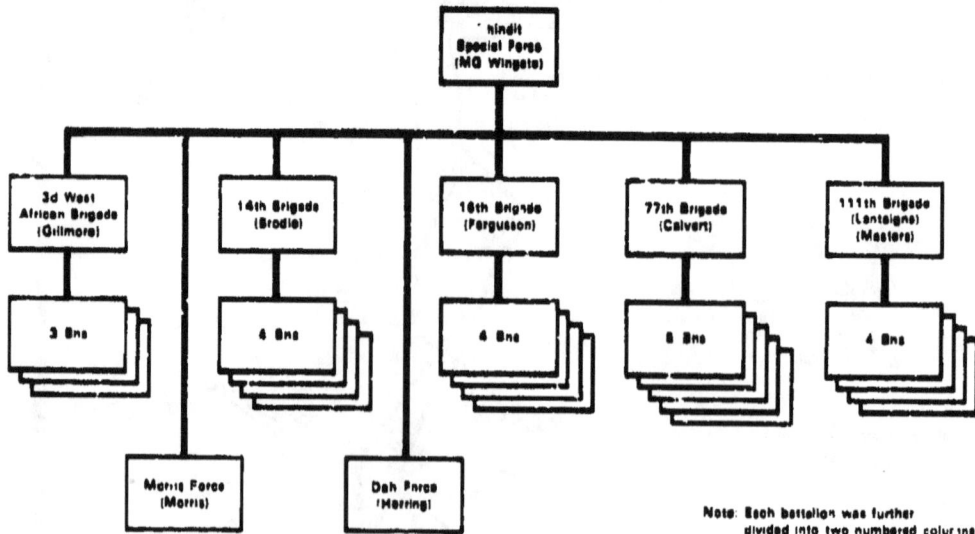

Figure 1. Chindit organization

It is noteworthy that one of the original (1943) Chindit battalions was the 13th Battalion King's Liverpool Regiment, a battalion of older men brought up in the urban manufacturing city of Liverpool. Despite their apparent unsuitability for jungle warfare, this battalion fought creditably, although they suffered a higher ratio of casualties than the other two original Chindit battalions.[9] Similarly, during Chindit II, the many other regular-line infantry battalions (the South Staffords, the Cameronians, the Leicesters, etc.), the Nigerian battalions, and the artillery and armored-car reconnaissance units that were converted into Chindit infantry, all showed a remarkable ability to adapt to the unusual, unorthodox requirements of the jungle. The key to their success was good, hard, relevant training. In fact, these diverse units were converted to Chindit infantry in just twenty weeks.[10]

The American Chindit brigade, Galahad, began forming shortly after the Quebec Conference in 1943. Eventually given the awkward (and hated) name of the 5307th Composite Unit (Provisional), this brigade comprised 950 men from various Pacific commands, 950 men from the Caribbean Command, and the remainder from stateside units. All were volunteers, and most had actual combat experience or training in jungle warfare. The men were organized into the 1st, 2d, and 3d Battalions and further subdivided, like the Special Force brigades, into two columns, which the Americans, however, chose to call combat teams (see figure 2 and table 1).

As noted earlier, Galahad was composed entirely of volunteers—men who signed on for an undefined mission of hazards and danger. However, these volunteers were unscreened and were not elite soldiers. They formed a wonderful mix of different types: dedicated professional soldiers, authors, intellectuals, criminals, students, and others.[11] If there were any unifying

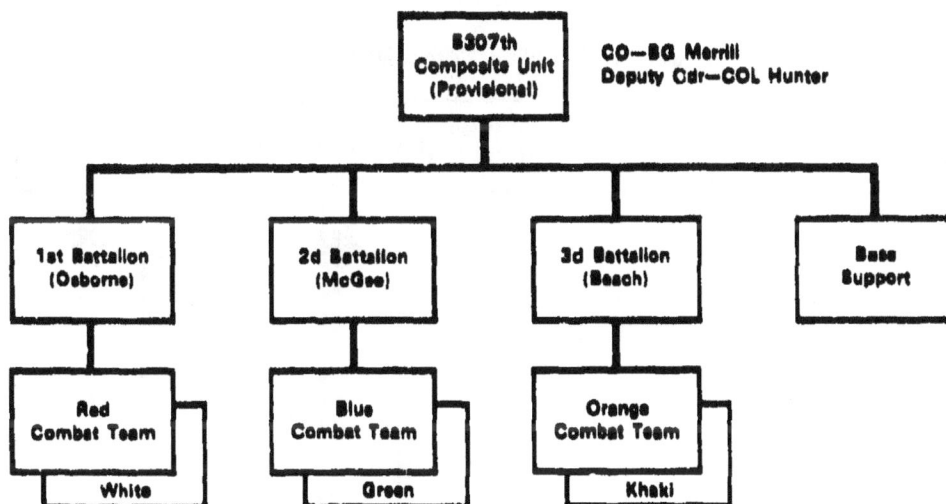

Figure 2 Organization of Galahad

14

characteristics, perhaps they were a common tactical background in jungle combat or training in jungle warfare and a wanderlust and desire for adventure and danger. The War Department had predicted an 85 percent casualty rate for Galahad before the unit was even formed for the operation.[12] Truly, the volunteers who knew this fact beforehand had undaunted spirits. In any event, the recruitment of the 5307th proceeded, and 3,000 men were quickly assembled and transported to India for organization and training as long-range penetration battalions.

All of the Chindit units were extremely light, being armed solely with small arms (rifles, pistols, light and medium machine guns) and light mortars. Once static blocks or strongholds were established, heavier weapons (40-mm antiaircraft guns, two-pounder antitank (AT) guns, 75-mm howitzers) were flown in for support. When on the march, however, the Chindits carried only what could be man- or mule-packed.

As a participant in the Quebec Conference, the U.S. Army Air Force commander, General "Hap" Arnold, wanted to provide the best possible support to Wingate. Accordingly, he directed the establishment of the No. 1 Air Commando under the command of Colonels Philip Cochran and J. R. Alison. Considered by some to be the most remarkable air fleet of the war, the No. 1 Air Commando comprised 13 C-47 Dakota cargo planes, 12 C-46 transports, 12 B-25 Mitchell bombers, 30 P-51 Mustang fighter-bombers, 100 light planes, 6 helicopters, and 225 Waco gliders.[13] The Air Commando was a temporary organization, however. It was scheduled to dissolve after ninety days, the time required to support the Special Force. The support of the Air Commando to the Special Force was essential; they could not have survived without it.

Table 1. Battalion Composition

	Battalion Headquarters	Combat Teams		Total
		No. 1	No. 2	
Officers	3	16	16	35
Enlisted men	13	456	459	928
Aggregate	16	472	475	963
Animals (horses and mules)	3	68	68	139
Carbines	6	86	89	181
Machine guns, heavy		3	4	7
Machine guns, light		2	4	6
Machine guns, sub	2	52	48	102
Mortars, 60-mm		4	6	10
Mortars, 81-mm		4	3	7
Pistols*		2	2	4
Rifles, Browning automatic		27	27	54
Rifles, M-1	8	306	310	624
Rockets		3	3	6

*NOTE: By the end of the campaign, many soldiers had acquired pistols as personal close-defense weapons of last resort

15

Training

Under the eyes of Wingate and his surviving commanders from Chindit I, both Galahad and the Special Force underwent the same kind of training regimen. There were two primary themes to the Chindit training. The first was physical endurance. One Chindit described the training program as a trial by ordeal.[14] The pace, duration, and intensity of the training were all designed to create and maintain an ultra-high level of stress and physical demands. Wingate's intent was to cut the deadwood early, to make the officers and men prove their ability to suffer and endure. One of the most celebrated Chindit commanders, Brigadier Michael Calvert, noted that three or four of his older commanders dropped out quickly, too old and unfit for the hardships. (They still, nonetheless, lent a helpful hand in the training.) Another brigade commander, John Masters, stated that no one over the age of thirty-five should have been permitted to remain in the organization; the physical stress simply was beyond their capability.[15] The Chindits were loaded with huge seventy-pound packs and marched unmercifully through man-killing jungle terrain.[16] No consideration was given to sickness, minor injury, heat, or weather. Placed on light rations and given little water, the men were pushed beyond the limits they thought they could endure. Such an approach was absolutely necessary. Without it, Chindit casualties would undoubtedly have been higher and effectiveness lower. On finishing their training, the Chindits were given time to recuperate and recover their strength before initiating actual operations. Regenerated, the Chindits crossed into Burma with high morale and supreme confidence.

The second theme of the training was "jungle craft"—a regimen through which Galahad and the Special Force received expert-level training in all the vital skills needed to operate behind the enemy's lines in trackless jungle. This included map reading, jungle navigation, scouting, patrolling, marksmanship, river crossings, watermanship, column marching, infiltration, night operations, terrain appreciation, squad, platoon, and company tactics, covering of tracks, evasion, and defensive operations. In particular, the men developed expert-level skills in map reading and land navigation.[17] Soldiers also trained for hand-to-hand and bayonet combat. In addition, Major General Wingate and Colonel Charles N. Hunter insisted on extensive cross-training: within Galahad, every soldier fired every weapon in the unit; platoon leaders and NCOs trained in artillery and mortar observation and in the use of the unit radios.[18] Should a machine gunner, mortarman, radio operator, or forward observer become a casualty, some other soldier was ready to take his place. Hunter placed his emphasis on platoon tactics, believing that in the jungle, every contact or operation eventually was decided on the basis of the effectiveness of the platoon. The Chindits also focused on individual decision making and initiative.

During the course of the training period, the Chindits also developed painstaking standing operating procedures (SOPs) for operations they frequently performed. Thus, SOPs covered such activities as performing river crossings, preparing landing zones for airdrops, establishing temporary harbors, initiating immediate actions on enemy contact, and establishing trail blocks.[19] Units

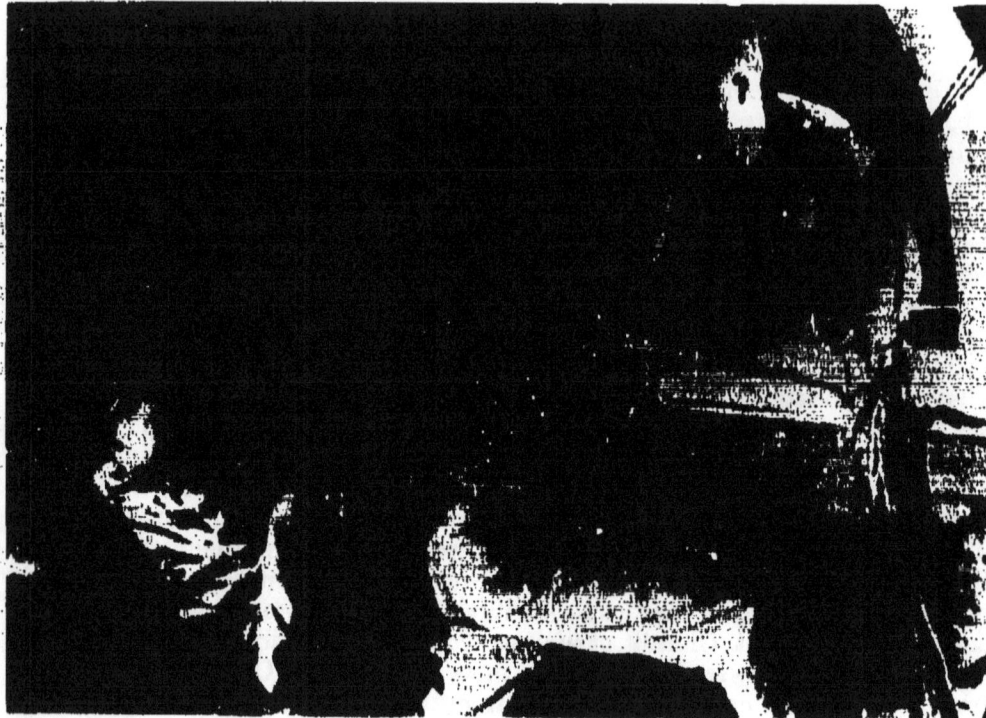

Kickers preparing to drop supplies in North Burma

had to be capable of executing these SOPs with clockwork precision and a minimum of orders. A few well-chosen words were sufficient to initiate a whole series of integrated actions based on individual tasks and teamwork.

Furthermore, units that planned to conduct specific kinds of operations received extra training in necessary skills. Thus, the two battalions of the 77th Brigade earmarked to establish a semipermanent block at White City spent fourteen days learning how to dig in deeply, establish overhead cover, emplace wire and minefields, and similar tasks.

The reconnaissance platoons also received a great deal of emphasis. Both Galahad and the Special Force used their best men in these units to provide critical functions regarding intelligence, warning, and surprise. The reconnaissance platoons were the elites of the Chindits, and they needed to be. Each Special Force column included a platoon from the Burma Rifles as their reconnaissance platoon. Calvert believed the Burma Rifles to be the best fighting unit in the Empire. Composed of men formerly residing in prewar Burma, the Burma Rifles knew the terrain, the people, and how to survive in the jungle better than any other regular battalion in the theater. Brave and devoted, the Burma Rifles possessed jungle skills that were exceeded only by the tribal Kachin levies who operated as guerrillas in the far north.

Another area that required special training was the handling of the pack teams. Because the Chindits depended so completely on their mules and horses to carry their heavy radios, ammunition, rations, and other vital supplies, it was absolutely essential that efficient handlers be trained to keep the columns moving. The veterinarians of the Special Force played an important role in this regard, providing good training and advice. Colonel Hunter was able to obtain enough Galahad volunteers with previous experience with mules or horses to take care of the 700 pack animals in the 5307th. Each Special Force brigade also utilized up to 1,000 mules.

Special attention was given to the very difficult training of the animals and their handlers for river crossings. Problems in this area, if not solved promptly, could hold a jungle column up for hours on a river bank. During the course of the operation, the men grew to love many of their animals, and they cared for and protected them just as if they were fellow soldiers. Incidentally, the Special Force debrayed their pack animals for security. Hunter, however, refused to debray Galahad's animals, stating that braying was one of the few pleasures a jackass enjoyed. He later claimed that to his knowledge their brays never posed a noise problem. Apparently, the mules were just too tired to bray.[20]

Training periods ended with lengthy field-training exercises under near-combat conditions. Wingate further stamped his influence here, frequently delivering scathing critiques to commanders and units that did not measure up to his exacting standards. Galahad participated with the Special Force in one of these ten-day exercises and held its own.

As a final training preparation before going into battle, Hunter hiked Galahad over 140 miles from their last assembly area to their jump-off site beyond Ledo. He claimed that this decision, despite its unpopularity, accomplished a number of goals. It completed the conditioning of the men and animals (who had lately undergone a soft three weeks of travel time by train). Next, it allowed the muleteers and their animals to adjust to each other on the trail. Furthermore, it "sweat in" the pack saddles to the animal's backs. And finally, it eliminated unfit men from the ranks. Hunter stated that, "More than any other single part of Galahad's training, the hike down the Ledo Road, in my professional judgment, paid the highest dividends."[21] Having completed this arduous hike with full loads, marching primarily at night over mountainous terrain, the men of Galahad lacked no confidence in their ability to meet the physical demands of the coming operation.

Operational Concept and Methods

The plans of Stilwell and Wingate basically were good plans based on a unique operational concept that originated more or less in Wingate's fertile mind. The fundamental revolutionary concept behind the Chindit style was Wingate's proven idea that a large, specifically trained, nonindigenous force could operate indefinitely in the enemy's rear. Wingate implemented this concept by using new technology to the fullest extent possible. Using the cargo aircraft as supply trucks, the radios as telephones, and P-51 fighter-bombers as direct-support artillery, Wingate was able to deploy, sustain, direct, and support the Chindits at a far greater depth than had appeared possible. The Chindits, in this sense, may be viewed as a prototype for later airmobile forces.

18

General Wingate on an inspection trip

Another fundamental feature of the Chindit style was its reversed system of values regarding terrain. Previously, Allied armies had considered the jungle to be their enemy. One of the feared traits of the Japanese was their ability to operate so well in the jungle. To the Chindits, however, the jungle was a refuge:

> Trails, which you had always thought of as friendly, were here the enemies, while the nighttime forest, almost the symbol of childhood terror, now meant blessed safety. You had only to lie quietly in the darkness of the forest and you were back in the invulnerable refuge of childhood's bed. No one could reach you without betraying himself with every step as he foundered among the myriad noise traps of leaves and branches.[22]

Furthermore, terrain that the Japanese believed to be impassable formed secret avenues of approach for the Chindits, who demonstrated an unparalleled capability for tactical and operational maneuver.

Operational-level maneuver characterized both the deployment and the subsequent operations of the Chindits. The secret, 450-mile-long march by the 16th Brigade over terrible terrain to the close approaches of Indaw is a prime example. Unfortunately, in this instance, the brigade forfeited its surprise and ability to concentrate by conducting a premature attack without adequate reconnaissance. Thus, it failed to secure its objectives in the Indaw area because a flawed tactical operation offset a brilliant operational maneuver.

The assault by Galahad in taking the Myitkyina airfield is another example of operation level maneuver. Already exhausted and decimated from having marched and fought through 500 miles of forbidding terrain, the

5307th, after eleven weeks of close, deadly encounters and constant stress, was approaching the end of its effectiveness when Stilwell ordered it to move 65 more miles through the jungle to attack Myitkyina from the north.[23] The line of march crossed the forbidding 6,000- to 8,000-foot-high Kumon Range. So formidable was this obstacle that the Japanese had not bothered to outpost it. Nonetheless, in an incredible feat of endurance and sheer perseverance, Galahad appeared undetected on the outskirts of Myitkyina after eighteen days. The nature of the march is described vividly by Charlton Ogburn:

> We set off with that what-the-hell-did-you-expect-anyway spirit that served the 5307th in place of morale, and I dare say served it better. Mere morale would never have carried us through the country we now had to cross. We had fought with mountains before, but none like those of the Kumon Ranges under the monsoon rains. . . .
>
> We were scarcely ever dry. When the rain stopped and the sun came out, evaporation would begin. The land steamed. The combination of heat and moisture was smothering. You had to fight through it. For those most weakened by disease, it was too much. For the first time you began to pass men fallen out beside the train, men who were not just complying with the demands of dysentery—we were used to that—but were sitting bent over their weapons, waiting for enough strength to return to take them another mile. During the worst times heretofore we could always count on one thing to keep us going—and that was the process of keeping going itself. As long as the column was on the march, men somehow seemed to be able to keep up, and it was only when we laid up for a day that the sufferers would collapse. But it did not work any longer. We had stragglers. Whenever we bivouacked, men who had been incapable of keeping up with the column, slowly as it moved, and were too tired to worry about the danger from any Japanese there might be lurking about, would be plodding in for hours afterward, unsmiling and clammy with sweat. There was a feeling in the organisation that it was coming apart. And Myitkyina was still 60 miles away.[44]

After a quick reconnaissance and a short rest period, Hunter captured the airfield in a quick daylight attack that caught the Japanese completely off guard. At the conclusion of this short battle, with the airfield in Allied hands, Hunter had only 1,310 Galahad soldiers left of 2,200 that had started the trek. Almost all of the casualties were of the nonbattle variety: injury, disease, and exhaustion. Virtually every soldier had a fever of some kind and was plagued with oozing sores or dysentery. Despite these terrible losses, the capture of the airfield was a stunning success, thanks to the maneuver that only the Chindits could have performed, coupled with a prompt, effective tactical attack.

The secret airborne deployments of the 111th and 77th Brigades into Chowringhee and Broadway also constituted operational-level maneuver. In just six days, from 5—10 March, Wingate inserted 9,000 men and 1,100 animals secretly into the enemy rear. The deployment also served a deceptive purpose by confusing the Japanese regarding the size and intent of the forces operating in their rear.

The establishment of large, permanent blocks on the Japanese main supply routes was a new concept for Chindit II made possible by the sevenfold increase in forces given to Wingate. The blocks themselves were the responsibility of specific brigades. Two blocks were established: one block, called White City, near Mawlu by the 77th Brigade and the other block, called

Allied casualties at Myitkyina airfield

Blackpool, which was placed farther north in the Railway valley near Hopin by the 111th Brigade. In each case, the block was established by two or more battalions that occupied static defensive positions that were well-fortified and dug-in. The battalions received additional supporting weapons by air. Outside the block, a mobile "floater" column or columns patrolled to be ready to attack any enemy from the rear or flank that tried to clear the block. These floater columns maintained radio contact with the block and were relieved periodically to keep them fresh. Floater columns, however, did not always work out well; sometimes they lacked the necessary punch to take on a strong, alert enemy force.

Similar to the block in organization, but serving a different purpose, was the stronghold. The idea of the stronghold was another key element in Wingate's operational concept for Chindit II. Its basic purposes were to serve as a fortified base, a port of entry for reinforcements, a shelter for recuperating columns, and a collection point for casualties. These strongholds, islands in a sea of jungle, are described in Wingate's training note:

> The Stronghold is a machan overlooking a kid tied up to entice the Japanese tiger.
> The Stronghold is an asylum for L.R.P.G. wounded.
> The Stronghold is a magazine of stores.
> The Stronghold is a defended airstrip.
> The Stronghold is an administrative centre for loyal inhabitants.
> The Stronghold is an orbit round which columns of the brigade circulate. It is suitably placed with reference to the main objective of the brigade.
> The Stronghold is a base for light planes operating with columns on the main objective.

21

> We wish, therefore, firstly to encounter the enemy in the open and preferably in ambushes laid by us, and secondly to induce him to attack us only in our defended Stronghold. Further, to make sure of our advantage, and in view of the fact that the enemy will be in superior force in our neighbourhood we shall choose for our Strongholds, areas inaccessible to wheeled transport. For convenience sake such Strongholds should clearly be used to cover (but not to include) an airstrip. The ideal situation for a Stronghold is the centre of a circle of thirty miles radius consisting of closely wooded and very broken country, only passable to pack transport owing to great natural obstacles, and capable only of slow improvement. This centre should ideally consist of a level upland with a cleared strip for Dakotas, a separate supply-dropping area, taxiways to the Stronghold, a neighbouring friendly village or two, and an inexhaustible and uncontaminatable water supply within the Stronghold.
>
> The motto of the Stronghold is "No Surrender."[25]

The two strongholds established by the Chindits were at Broadway and Aberdeen. Broadway attracted a large Japanese force on itself, a force it defeated while remaining fully in operation. Aberdeen was never attacked by an enemy ground force. Although there were problems with both sites—primarily their inaccessibility—the concept of the stronghold turned out to be feasible and useful in practice. Interestingly, the White City block served some of the same functions as the remote strongholds in that fresh reinforcements were delivered directly to White City and many casualties flown out.

Chindit Tactics

Clearly, the implementation of the operational concept of the Chindits required them to display unusual, specialized light infantry tactics. As the primary author of Chindit tactics, Wingate deserves much credit, although the contributions of Calvert, Fergusson, and others—including the influence of the Japanese themselves—should not be overlooked. The long-range penetration tactics used during Chindit I were essentially evasive. Success depended on the superior tactical mobility of the Chindits and on high levels of jungle craft. During the first expedition, the Chindits were a well-armed, cohesive, hit-and-run force. Emerging from the jungle, they struck a poorly defended target, destroyed it, and then faded away into the jungle, drawing the enemy after them in fruitless efforts at retaliation. Over an extended period of time, the Chindits conducted continuous raiding, always striking the enemy by surprise where he was weakest and then slipping into the safety of jungle sanctuaries. Although their operations were guerrilla-like, the Chindits did not practice guerrilla warfare in the classic sense. Unlike guerillas, the Chindits were better armed, and they hit the Japanese with a harder punch and with higher frequency than would guerrillas. In addition, the Chindits were sustained primarily by airdrops of supplies, not from indigenous sources. Also, Chindit operations, although decentralized, submitted to directions from a Special Force headquarters; thus, their operations lacked the haphazard nature of most guerrilla operations.

Some of the Chindit elements continued to act in the classic long-range penetration group (LRPG) style in 1944. Perhaps the best example was the Morris Force, which operated against Japanese lines of communication situated along the Bhamo-Myitkyina road. Because the Morris Force was too

Action at Broadway

weak to establish a permanent block, it had to perform a harassing, inter-
dictory role instead. The mountains just to the east of the main supply route
(MSR) provided an excellent refuge for these marauding columns. The effectiveness of the operations of the Morris Force and the Dah Force (another
small British element led by Colonel Herring that used Kachin tribesmen for
raiding in the same area) is described clearly by Brigadier Shelford Bidwell:

> The three Gurkha columns (of the Morris Force) between them . . . had
> demolished eight large bridges, including two iron ones, two ferry installations
> and numerous small bridges, and had blown a long section of the road from a
> cliff face into the gorge below. The Gurkhas had come down from their moun-
> tain fastnesses to hit the road in six different places, and the two forces
> between them, by means of patrols, ambushes and attacks on the road engi-
> neers, prevented the repair of the breaks. Supplies to Myitkyina were never
> completely cut off, but they were substantially reduced. Units the Japanese
> could ill spare were deflected to guarding the road and chasing Kachins (the
> Dah Force) and Gurkhas back into the jungle; all supply convoys were liable to
> ambush and the movement of all but large bodies of troops made hazardous.
> For a battalion and a half (the Morris Force) and Herring's little mission (the
> Dah Force) it was a handsome dividend, and a vindication of Wingate's purer,
> earlier doctrine.[26]

Initially, the 111th Brigade, before it was ordered north to establish the Black
pool block at Hopin, also pursued the same kind of tactics in its target area
northwest of Indaw and with similar excellent results.[27]

Lieutenant General Joseph W. Stilwell awarding medals at Myitkyina. Colonel Charles N. Hunter, second from left, had to borrow a shirt to complete the ceremony.

However, Wingate's ideas for the employment of the Special Force during Operation Thursday changed considerably from 1943 to 1944. With a much stronger force at his disposal, Wingate chose to modify LRPG tactics. Operations by the columns assumed a more direct, more aggressive character. Through the concepts of the block and the stronghold, Wingate elected to practice more or less conventional attacks and set-piece defensive battles, while the floater columns practiced the pure LRPG style. Implementing the new idea demanded that the Chindits switch from being guerrilla fighters to conventional warriors from one week to the next. The training for Chindit II reflected this concept to some degree. Calvert, understanding Wingate's new doctrine better than anyone else, clearly describes the new tactics:

> The main principle on which the Long Range Penetration (L.R.P.) Brigade was based was above all versatility. Versatility of maneuvre due to air supply and air casualty clearance. Versatility of power in that such a brigade could penetrate through every type of country in eight columns of about 400 men each, like the fingers of one's hand, and then concentrate in bringing the fingers together to clutch at the throat of the enemy when his attention had been duly scattered, or so strike a blow with a clenched fist at an important objective. When the brigade was concentrated in battle it re-formed into a more normal brigade of three or four battalions reinforced by artillery, heavy mortars, hospitals, engineer stores, etc., brought in by air. This turned it from a series of marauding columns into a homogeneous, co-ordinated brigade. Above all we placed our reliance on air.[18]

Not everyone else, however, was comfortable with the obvious conflict between evasive, hit-and-run LRPG tactics and that of the stand-up fights required when holding a block or assaulting a strong position. Brigadier Fergusson, for instance, still favored the old Chindit style, even when directed onto a substantial target. In the attack against the enemy at Indaw,

24

Fergusson's plan, "using widely dispersed columns coming in on the objective from different directions, had a strong Chindit and guerrilla flavor."[29]

In his study of the Chindit War, Brigadier Bidwell concluded:

> "There is a marked contrast between the operations of Calvert's brigade and Fergusson's. The difference was not of skill, but of style. . . . Calvert had veered more and more toward the conventional. He closely controlled his columns, had trained them to dig and fortify positions and was fully prepared to assault the Japanese, controlling his vital air support through his brigade Royal Air Force officer. The 16th Brigade represented an earlier phase in the evolution of Wingate's tactics. Fergusson commanded loosely, leaving the details to his widely dispersed column commanders."[30]

Tactical Movement

Regardless of whether the ultimate objectives were fleeting targets or ground to be secured and held, all the Chindit brigades used the same method of tactical movement in the jungle. As previously mentioned, the column formed the basic tactical element. Generally, the Special Force column consisted of a strong infantry company of four platoons, an engineer-commando platoon skilled in booby traps and demolitions, a heavy weapons platoon (two medium machine guns and two three-inch mortars), a reconnaissance platoon (mostly Burma Rifles), an animal transport platoon, and the headquarters platoon, which included an RAF detachment, intelligence section, signal element, and medical detachment. The combat teams of Galahad were almost identical in form except that they were a little stronger. Each combat team was based on one and one-half rifle companies. The heavy weapons platoon included three or four heavy machine guns, four or six 60-mm mortars, and three or four 81-mm mortars.[31] Each Chindit battalion was divided into two columns commanded by the battalion commander or his second-in-command/executive officer. Apparently, company operations and normal company command were not typical features of Chindit warfare.

When on the march, the Chindits moved fast. Galahad particularly was noted as being extremely fast at covering ground in the jungle. While moving, the reconnaissance platoons always preceded the main body, sometimes by as much as several miles. Trail column formations most often had a rifle platoon as the first element in the main body, followed by a rifle company with half of the heavy weapons platoon. Column headquarters, transport, and the medical detachment formed the middle of the formation, with another rifle platoon and the rest of the heavy weapons in the rear.[32] Some columns strictly separated the combat elements from the support elements so that there would be no delay in the tactical deployment of the column's combat power.

The column usually traveled, on and off trails, in a long, single file. When several columns moved together, the entire formation could stretch a long distance. While the accompanying mules slowed the columns, they were absolutely essential. Depending on the terrain, a day's march might vary anywhere from a couple of miles to fifteen miles. During most of the marches, soldiers suffered mind-numbing exhaustion, which forced them to concentrate all their energies on the need to keep moving. Often, the mountainous trails were negotiated on all fours. When the mules were unable to climb, their burdens were unloaded and man packed to the top of grades.[33] As the units suffered losses, particularly of animals, agonizing decisions had to be made about what to carry on and what to leave behind. To maintain the pace, men had to have extreme levels of physical endurance and discipline. One ex-Chindit noted that during his three-month campaign, his column was always exhausted before battles.[34] The men of the 2d Battalion of Galahad were so exhausted after making a forced march into a defensive perimeter at Nphum Ga that some of them actually slept during the ensuing enemy barrage and ground attack.[35]

In addition to the security provided by the reconnaissance platoons, columns often used smaller elements on connecting trails, either to provide early warning or to block an enemy infiltration or patrol. To avoid detection, Galahad worked down river valleys. During the movement to Nphum Ga, one of its combat teams crossed a river forty times in one day.[36]

An unexpected contact on the march, which often consisted of a Japanese trail block or ambush, prompted immediate action. In response to Japanese fire, the point squad of the leading rifle platoon immediately took to ground and established a base of fire. The following two squads moved off the trail to the left and right and worked through the jungle to attack the flank or rear of the enemy position. Once the Japanese felt this flank pressure, they normally abandoned their block only to set it up again farther down the trail. These tactics reduced casualties but took a lot of time. If the Japanese position was particularly strong, the Chindit commander called up his heavy weapons to blast the enemy, while stronger elements moved off the trail to clear the block.

When attempting to hold off the advance of a Japanese unit, the Chindits used Japanese tactics in reverse. In one case, two U.S. platoons (90 men) held off a Japanese force of 850, while the main bodies of 2 battalions cleared a trail to their rear. Withdrawing through successive trail blocks, established by each platoon in turn, these Galahads inflicted sixty enemy casualties without suffering a single loss of their own.[37] Machine guns, sited for mutual support, formed the basis of these stiff trail blocks.

The Attack and Defense

The typical Chindit attack involved a tactical march off the trail to the near approaches of the objective. Then, a quick, professional reconnaissance by the assault elements collected critical information regarding the enemy's strength, disposition, and level of alertness. Finally, undetected, the Chindits crept as closely as possible to a flank or the rear of the enemy positions. Having achieved surprise, the Chindits then attacked with a high volume of fire from two or more directions while being supported by well-sited machine guns and mortars. The shock effect of such an attack usually was sufficient to drive off the defenders, even if the attackers were outnumbered. The key element of such an attack was the use of surprise, envelopment tactics, well-aimed and well-disciplined fire, and shock.[38] Shock was enhanced because up to 90 percent of the Chindit columns were actual fighters.[39] If ground was to be held, either permanently or temporarily, the Chindits immediately went

over to a stiff defense to fight off the inevitable Japanese counterattacks. Most Chindit attacks were mounted in the daytime; nighttime attacks occurred infrequently.

Japanese counterattacks often took hours in coming, but when they came, they were ferocious and unrelenting. Galahad and the Special Force learned to dig-in deeply and rapidly. Overhead cover was particularly important, because the Japanese could often bring up artillery and even tanks—two weapon systems the Chindits did not have. The Chindits quickly stockpiled their machine gun and mortar ammunition for ready access. Heavy weapons were also dug in or revetted. Animals normally were moved to the center of the perimeter and were also revetted. The men also dug trenches for wounded personnel and established a medical treatment area in a protected position. Commanders always took care to ensure that the defensive position included an uncontaminated water source. In addition, paths were tramped out to listening-observation posts to reduce noise and to make night movement easier.

When the Japanese were heard assembling for an attack, the Chindits prepared their assembly areas with mortars. To support the Chindits, P-51 Mustangs of the No. 1 Air Commando attacked the Japanese at least twice a day depending on the weather, fulfilling the role of Chindit artillery. The Chindits learned to trust these airborne gunners implicitly. (The Royal Air Force and U.S. Army Air Forces detachments used procedures to contact the Chindits and to mark targets that are still applicable today.)[40]

27

American Nisei also enhanced the defense. Listening carefully to orders from enemy officers, the Nisei translated them in time for the defense to react. At Walawbum, a Nisei tapped directly into a Japanese wire line. The Niseis also confused the Japanese by shouting contradictory orders to induce them to charge. The Niseis were so valuable that Stilwell offered one to the British so that his talents could be used at White City.

Above all else, however, marksmanship and fire discipline were the keys to Chindit defense, particularly for Galahad. The accurate fire of the American marksmen always took a heavy toll of the Japanese, who were poor shots with individual weapons. Repeatedly, well-placed rifleman picked off Japanese while they m___ed carelessly forward. So confident were the Galahads in their fire that they routinely let the Japanese approach to grenade-throwing distance before unleashing, on signal, a devastating volume of fire that mowed down the charging enemy in rows. Colonel Hunter (and others) stated unequivocally that superior American marksmanship was the single most important feature in Galahad's campaign.[41] Conservation of fire was also important, for the Chindits were usually surrounded by the Japanese and had to depend on airdrops for resupply of ammunition. Thus, the Chindits were careful not to let their stocks dwindle, especially the mortar and machine-gun ammunition. British and Americans alike have stated in their memoirs that the machine guns and mortars played an absolutely essential role in Chindit defensive operations.

In the course of its operations, Galahad normally was relieved from its defensive roles by the arrival of Chinese regiments.[42] The British, however, established permanent blocks at White City and Blackpool and at the strongholds of Broadway and Aberdeen. These positions, naturally more fixed in nature, encompassed airstrips within their defensive schemes and included, in the case of Broadway, a temporary fighter-bomber element on station. The Chindits strengthened these positions with wire and booby traps, constructed communications trenches and thick overhead cover, brought in more firepower in the form of antiaircraft, antitank, and artillery weapons, and replenished the troop garrison.

The most unique feature of the defense of the permanent sites, however, was the integrated use of floater columns and "jitter" patrols. Calvert, in particular, used these mobile elements outside the wire to ambush the Japanese, striking them in the rear. The Japanese proved very vulnerable to attacks by unknown forces against their rear, such assaults often causing them to call off their own attacks. In this way, the defenders were able to retain the initiative until the Japanese moved in vastly superior forces.

In the long run, the Chindits were ill suited to conduct or to withstand a long siege. At the end of a more than 200-mile-long air line of communication, they simply did not have the combined arms combat power for these tasks. In addition, no replacements existed to replace Chindit casualties. Nevertheless, their efforts tied up large numbers of Japanese forces and caused huge Japanese losses. In Galahad's first defensive block at Walawbum, it is estimated that the Japanese lost 800 men to Galahad's 45 killed and wounded in action.[43] Eight battalions of Japanese broke against the reefs of the White City Brigade, and an entire Japanese regiment was decimated at Blackpool before the 111th was forced to abandon its block.[44] Initially shocked, then infuriated by the stubborn defenses they encountered, the Japanese soon became demoralized by their losses. Had they been more patient and employed heavier forces, particularly artillery, the Chindit blocks could have been cleared away with fewer losses. However, engaged as they were in the decisive battle of Imphal-Kohima, the Japanese were loath to divert their best units and their scarce equipment to clear the blocks.

Kachin Support

The final notable feature of Chindit tactics was their reliance on support from the North Kachin levies. Raised initially by the British, led by Special Operations Executive (SOE) and OSS officers, and aided by the ever-ministering C-47s, the Kachins initially conducted their own guerrilla war of spying and sabotage against the Japanese—often with spectacular results.[45] In 1944, however, their operations were consciously coordinated with those of the Chindits.

The Kachins were incredibly light, being armed with a variety of new and ancient weapons, but they carried little ammunition. They avoided casualties by using hit-and-run tactics and by choosing the time and place of contact. The Kachins were also jungle masters with a near telepathic ability to regroup after dispersal. Noiseless, adept at night, brave, heroic, and possessed of a fierce fighting character, one writer described them as the greatest fighting men in the world.[46] Every British and American commander who enjoyed their support praised their remarkable fighting ability and jungle craft.

A squadron of Kachin Rangers preparing for inspection

Kachin support to the Chindits took several forms. In the main, the Kachins furnished intelligence, acted as guides, screened Chindit movements, and provided trail security. In a few instances, they reinforced the Chindit columns. The Chindits also borrowed Kachin elephants to carry supplies and clear drop zones. The Kachins even built footbridges across rivers and improved trails to assist the progress of the columns. The 5307th used their special skills more than the Special Force, because the 5307th operated in the northern areas where the Kachins were most numerous. After the fall of Myitkyina, Hunter remarked in a cable to the commander of OSS Detachment 101, that the 5307th could not have succeeded in its tasks without the help of the Kachins.[47] In short, these brave, loyal, primitive tribesmen proved to be indispensable to the Chindits. They represent an excellent example of light infantry forces making use of indigenous resources and cooperating with local irregulars.

Logistics

Chindit logistics depended entirely on two means: airplanes, to deliver supplies deep in enemy territory; and mules and horses, to haul them once they were received. Everything the Chindits needed came by air. Without the services of the No. 1 Air Commando and the 2d Troop Carrier Command (which supplied the 5307th), no Chindit expedition would have been possible. The concept of long-range penetration was tied directly to air supply. Chindit I had proven that air supply was feasible; Chindit II stretched the concept to its limits and gave new meaning to the scale of deep penetration.

Ingenuity, innovation, and energy formed the basis for the Chindit air lifeline. Air operations required good, reliable communications; expert liaison; fast responses to Chindit requests; well-executed SOPs; fighter protection; and bold, skilled pilots. Teamwork and mutual trust were central to the achievement of success. Ground and air elements understood each other, and few misunderstandings existed. Some of the highest praise expressed by Chindit commanders concerned the heroic efforts of their air support.

The Chindits called for air resupply approximately every four to five days. This low frequency of resupply and the short duration of loiter time by air units over the drop zones greatly contributed to the maintenance of secrecy regarding Chindit locations. The ground elements normally transmitted their specific requests the day or night before the drop was made. Supply personnel in the rear packaged the material in such ways that some of it could free-fall to ground on a low-altitude pass, and the rest could descend by parachute.[48] All containers were configured to conform as much as possible to packboard loads, with little necessary resorting. Most supplies fell into common categories—food, ammunition, medicine, clothing, grain, and engineer stores. When requested, however, the materials delivered could be unique and personal:

> There was also a "personal service," which periodically dropped to individuals items they had stored with the supply officer before leaving, and which handled special requests as they arose. The R.A.F. made a valiant attempt to give the Chindits anything they asked for. Among the "personal service" items that traveled the aerial supply route were monocles, a kilt, false teeth, spectacles, pipe tobacco, boxes of snuff, small food luxuries, new books, notice to one man of an 11,000 rupee legacy, and one officer's last will and testament. Every

Chindit who had false teeth or wore glasses had left a plate impression or an eyeglass prescription on file at the air base. At every dropping, Wingate's men received mail from home, newspapers and magazines.[49]

Those who selected routes during tactical movement considered the proximity of good drop zones (which the jungle provided in good measure) and landing zones for heavy equipment. Landing zones were natural clearings, rice paddies, sand bars—even the surface of Lake Indawgi for pontoon-equipped aircraft. Occasionally, the Chindits had to chop drop zones out of the jungle, but this often could be done quickly, particularly when Kachin elephants were on hand to help. Supplies received at drop zones were promptly loaded and moved away from the area, distribution normally being made once the column had reached a secure position.

The support base also demonstrated that it could be very responsive when necessary. Although the average time between request and delivery ranged from twelve to eighteen hours, the shortest time for a supply mission to reach Galahad was a mere two hours and twenty-two minutes after the message had been received.[50] The Galahad support base monitored the operation's radio net in order to keep track of the unit's whereabouts and needs. This practice eliminated the need for retransmission of messages. In addition, some trucks and aircraft were preloaded, thus ready to dash to the airfield as soon as a supply request was received.

Most of the supplies for the permanent Special Force positions, including all the heavy gear, was airlanded. Accordingly, the airstrips within the blocks and strongholds had all been made usable for C-47s by airfield engineers. When these airstrips became unusable by reason of Japanese interdiction, cargo aircraft delivered their loads directly to the defensive positions by parachute. This practice, in fact, gave White City its name, the trees in the perimeter having been covered with white parachutes from supply drops. (Incidentally, the parachute silk made good trading material to local villagers in return for fresh food.)

One of the most important features of the air lifeline was medical evacuation. During Chindit I, no allowances had been made to evacuate casualties. Thus, a serious wound or injury usually meant the death or capture of the victim. During Chindit II, however, casualties were evacuated on the L-4 and L-5 light planes, often within mere hours of the wounds. The American pilots of these planes proved their ability and willingness to land almost anywhere to pick up Chindit casualties, which created a strong bond of admiration and affection between the ground and air forces. The knowledge that this capability existed had an immeasurably positive effect on Chindit spirits. Air evacuation probably had more influence on the maintenance of good morale than any other facet of the organization.

The significance of the air support to the Chindits cannot be overemphasized. No other means of supply could have sustained the force. Moreover, every time that the Chindits found themselves in a tight spot—such as at Nphum Ga and at White City—airdrops kept them in the fight. On the other hand, when Japanese fire closed the Blackpool airstrip, restricting resupply to airdrop alone, the 111th Brigade was forced to abandon its block, partly because it could not be sustained in place. This failure demonstrated the absolute necessity of the air lifeline.

The mule packtrain was the other basic element of Chindit logistics. The two most important loads carried by the mules were the heavy radio sets and the crew-served weapons and ammunition. With the packtrains, the Chindits were able to deliver a much harder punch than the classical, historical guerrilla, and it gave them substantial staying power in the more protracted battles. Furthermore, both Galahad and the Special Force used the animals as mounts for leaders.

The Chindits did make some effort to use indigenous supplies when they could. The Special Force carried silver rupees to buy food and information from the Burmans and native tribes. During its battles near Shaduzup, one Galahad outfit, after running off the enemy, ate the rice and fish breakfasts the enemy had prepared and then changed into the fresh underwear they found on a supply truck.[51] Some consumption of natural jungle foods also occurred. On the whole, however, the Chindits depended on their air lifeline for the bulk of their supplies.

The food provided the Chindits during the war is a sore point to most surviving Chindits.[52] Galahad subsisted on the following rations: 80 percent K rations, 5 percent C rations, 5 percent 10-in-1 rations, and 10 percent B rations. The K rations were survival rations intended to keep men alive but not to sustain them, especially for the arduous campaign endured by the 5307th. K rations were used because they were lighter, they did not spoil, and they were easy to supply. But these rations lacked bulk and energy. Constant reliance on K rations ultimately produced exhaustion in the men and caused their stomachs to shrink to the point where they could no longer tolerate fresh solid food during the rare times it was available. These effects were easy to predict. They were, in fact, accepted on the expectation that Galahad would be in the field no more than ninety days. But this was an expectation that was not met. The Special Force also relied almost entirely on K rations, but their diets were supplemented with an occasional issue of bully beef or some other more filling fare. In retrospect, reliance on K rations was a grave mistake. More attention could have and should have been given to a more nutritious food supply.[53]

Leadership and Morale

High levels of morale and esprit were developed in the Special Force and Galahad. These were generated, in part, by the arduous training these units endured. The skills they acquired in training produced in the men supreme confidence in their abilities to succeed. Further improving their morale was their assurance that they would be evacuated by air should they be wounded or hurt. Morale was also strengthened by the Chindits' trust in their capable tactical leaders: the men were willing to place their lives in their leaders' hands. Finally, the Chindits' self-confidence was reinforced by their knowledge that they were participating in a unique and dangerous operation that they alone were fit to conduct. Beyond these common factors, however, there were other marked contrasts between Galahad and the Special Force regarding the quality of their leadership and morale.

Troops of Merrill's Marauders resting on a mountain trail

The British soldiers, on their part, enjoyed the special cohesion inherent in the individual regiments, a cohesion based on a common heritage, personal friendships, and a common racial, social, and ethnic background. Furthermore, the British brigades had a strong personal faith in the abilities of Wingate. They were also convinced that he was concerned about their care. As long as Wingate was at the helm, the Chindits believed that the good ship Special Force might transit rough water, but it would always keep sailing. Wingate's premature death three weeks into the operation shook them.

The British officers of Wingate's command, however, continued his high standards. They were men who traditionally led by the force of example. Their personal bravery in leading charges, in willingly exposing themselves to fire, and in remaining calm during moments of rising panic and terror strongly impressed their men and produced some of the most exhilarating examples of courage in the war. Calvert, in particular, gained the reputation as one of the most courageous warriors in the theater.

Several other leadership techniques of Calvert bear mentioning. For one thing, he rejected the idea that the Chindits were survivalists living at risk in a hostile sea of Japanese. Instead, he instilled in his men the idea that the Chindits were the kings of the jungle, who would boldly seek out the

33

Japanese in their lairs.[54] Let the Japanese come, he preached, if they were brave enough to meet their doom. Calvert also directed that each officer and noncommissioned officer account for every man in his command, alive or dead, in every action. Moreover, his officers were to take pains to explain new plans to their men, reviewing dangers calmly, in an effort to relieve their apprehensions. Understandably, these measures instilled confidence and kept morale high.

In contrast, Colonel Morris, in moving his column tentatively and over-cautiously, degraded the morale of his men. Not liking to take risks, he was never comfortable with the hit-and-run tactics of Wingate. The haphazard nature of Chindit operations disturbed him. The idea that the enemy could be anywhere caused him much apprehension. He communicated this uncertainty to his command, increasing its fears.[55]

But the Special Force was well looked after in a number of other ways. "To prevent unnecessary anxiety among the relatives of the Chindits, a special airgraph (air message) service was organized. At regular intervals, each man's family and close friends were notified whether he was alive and well. In addition, every man had made out a list of special dates he wanted remembered—birthdays, anniversaries, and the like—and as each date fell due, Chindit

The 2d Battalion crossing the Tanai River over a native bridge on the way to Inkangahtawng, March 1944

headquarters sent the appropriate telegrams."[56] The personal service of the Special Force support base has already been described; the laudatory effect on morale of the periodic luxury drops of comfort items—cigarettes, rum, chocolate candy—bears reiteration. Furthermore, some mail was delivered in every supply run and in every flight of the evacuation aircraft. The British also conscientiously awarded in-field promotions and decorations to deserving soldiers. In short, even though isolated from their parent command by 200 miles of enemy territory, the Special Force did not feel abandoned. Strong links to the main army, to family, and to the home island were maintained.

Galahad's situation, on the other hand, took on a reverse image of that described above. Plagued from the start with the awkward title of the 5307th Composite Unit (Provisional), Galahad had no history, no colors, no patch, and no crest. When they designed their own crest, it was bureaucratically rejected. As a result, these diverse men gathered in from all over the world for a temporary mission had no symbol of unity around which to cohere.

Nor did they have a charismatic figure like Wingate on which to focus. Colonel Hunter trained them and led them as de facto commander until Brigadier General Frank Merrill assumed command on 4 January 1944. Even then, Merrill's weak heart and his reliance on Hunter reduced his influence within the unit. His field leadership also was suspect, especially after he suffered two heart attacks and was evacuated twice during an operation. Hunter was the true commander of the unit in all but title. As for Stilwell, the soldiers of Galahad had no affection for him. In fact, many came to hate him for his callous treatment of the unit and for his bewildering refusal to recognize their contributions. Stilwell, unfortunately, lavished all of his attention on his Chinese divisions and all but ignored the valiant men of Galahad.[57]

Despite these handicaps, however, the 5307th was molded into a fearsome instrument of war. Brigadier Bidwell gave them very high marks, describing them as infinitely adaptable.[58] Another British participant in the campaign thought they were laconic and unemotional, perhaps "the most professional" of all the Chindit groups.[59] Any thoroughbred, however, can be bruised with rough handling; such was the case with Galahad.

Faithfully performing every tactical mission given them, Galahad received no comfort or luxury supplies and almost no mail. One unit in the 5307th went two months without mail.[60] Moreover, unit officers and men received no decorations until after they had captured the Myitkyina airfield (and then only sparingly) and no promotions at all until they were withdrawn from the area. Inquiries into this matter were received with disdain and scorn.[61] Stilwell visited them several times in the forward area, but only after the capture of the Myitkyina airfield. Even then, he made no attempt to greet the men or assess their condition. Without a doubt, this kind of treatment produced widespread feelings that the unit was a bastard organization, unloved and unrecognized. Years afterward, the thoughts of this abysmal leadership rankled the memories of the survivors.[62] No heritage, no colors, no crest, infrequent mail, no decorations, no promotions, no comfort supplies, no recognition: it is a wonder and a cause for admiration that Galahad performed as well as it did.

Operational Leadership

At a higher level, the senior commanders of the Special Force and Galahad failed their men in an operational sense. Wingate failed, in the first instance, by not focusing the Special Force on a single operational goal. He employed the force piecemeal and frequently changed their objectives. Slim, by his own admission, failed to see the flaw in Wingate's deployments. When he sent in the 14th Brigade and the 3d West African Brigade (after Wingate's death), he ordered them to be flown in to operate in the north where they were no longer needed rather than to the west. "I was wrong. I should have concentrated all available strength at the decisive point, Imphal. I fell into the same error as so many Japanese commanders. I persisted in a plan that should have been changed."[63]

Stilwell and his staff also committed serious operational failures, two of particular moment. The first occurred as the 5307th was in the process of blocking the Japanese lines of communication near Shaduzup. Alerted that the Japanese were trying to outflank his main formations in force by moving on an unguarded jungle approach, Stilwell's staff ordered Galahad to move by forced march and to take up a fixed defensive position at Nphum Ga to block the Japanese. Hunter characterized this change as "a new role for Galahad, one not contemplated when it was organized."[64] An official Army history described the mission as a radical change in concept. As a result, the 2d Battalion spent 11 days in a 400-meter-long defensive perimeter under almost constant artillery attack and ground pressure from the Japanese, while the 3d and 1st Battalions struggled manfully to relieve it. In the end, the Japanese withdrew—Galahad had won—but the fighting edge of Stilwell's most obedient and mobile troops had been worn dull.[65]

Stilwell's second error was his failure to take advantage of Galahad's coup de main at Myitkyina airfield. Despite having directed Galahad to the objective, Stilwell apparently had no well-thought-out plans on what to do after the airfield was in his hands. This mental lapse enabled the Japanese to build up the Myitkyina garrison to the point where it could only be taken after a three-month siege, not by storm. Stilwell's error nullified Galahad's heroic effort.

Tragically, Stilwell and the senior leadership failed to comprehend the full nature of the type of war in which they were engaged. Stilwell neither understood his men's frightening hardships nor their limitations.[66] The most obvious product of Stilwell's misunderstanding was his refusal to observe the ninety-day limit on the employment of the Chindits. Instead, Stilwell insisted that all the Chindits stay in the fight as long as there were men to bear arms. As a result, several of the brigades saw their strength fall to a fraction of their original strength. Naturally, as the individual Chindits became aware that the ninety-day limit was not going to he observed, their morale fell sharply. Stilwell's misunderstanding of the Chindit War also led him to misuse the Chindits grossly. For example, he assigned them conventional tactical missions far beyond their capabilities, the best example of which (and there are many to choose from) was the order to Calvert's 77th Brigade to take the fortified town of Mogaung in a frontal attack. Calvert succeeded, but he virtually had to sacrifice his command to do so.

Galahad had no one to protect it from Stilwell. By a twist of fate, the same was true of the Special Force. Not long after Wingate's death, Slim assigned the Special Force directly to Stilwell. Wingate's successor, Brigadier Lentaigne, however, had neither Wingate's stature nor his will. Consequently, he proved incapable of standing up to Stilwell to protect the Special Force from his deadly intentions. John Masters, acting commander of the 111th Brigade, described the failure of the high command to realize the strain that warfare in the enemy's rear had on the Chindits as the outstanding mistake of the campaign.

Costs and Problems

The most serious problem of the Chindit War was the misuse of forces. While Galahad was used properly in early operations, at the defensive battle of Nphum Ga, it was employed for tasks inappropriate to its training and capability. Later, at the long siege at Myitkyina, despite its heavy losses and exhaustion, Galahad was again rudely used in static defensive and offensive roles against the fortified Japanese garrison—roles for which it simply was not suited.

The Special Force, on the other hand, suffered from the start from an operational concept that consciously included a mix of conventional and unorthodox guerrilla tactics. While the Special Force demonstrated beyond doubt that it could perform both styles of warfare when required, the point is that when it used conventional tactics (à la White City), it failed to make maximum use of the special Chindit skills inherent in long-range penetration tactics. Any good infantry regiment could have held the White City block. But few regular-line infantry units could have moved through the jungle with Chindit speed and secrecy, struck the enemy with Chindit shock, or faded away with Chindit evasiveness. Holding territory was not a proper Chindit mission, because it did not take advantage of the Chindits' unique strengths.

The worst examples of misuse of the Special Force came at the hands of Stilwell, who continuously employed it as ordinary infantry. In addition to the previously mentioned attack on Mogaung by the 77th Brigade, one can also cite the urgings of Brigadier General Boatner, Stilwell's chief of staff, to the Morris Force to have its decimated force of Chindits and Kachin irregulars assault Myitkyina, a task impossibly beyond its capability. Brigadier Morris deserves only praise for his refusal to accept this mission from Boatner.[67] The horrible experience of the 111th Brigade at Blackpool is yet another example of misuse of the force. In this case, the blocking site was too shallow, too close to the front. Bombarded almost ceaselessly by artillery and within reach of heavy Japanese reinforcements, the 111th lay mercilessly exposed to the pounding of a much superior enemy force. With no choice other than obliteration, the 111th abandoned the position, much to Stilwell's disappointment.

Unfortunately, Stilwell's mishandling of the Chindits occurred at the worst possible time—during the last part of the campaign when the Chindits were already weakened by disease, exhaustion, and combat losses. Untrained and unequipped for such tasks, the Chindits found themselves ordered into battle when they were already on the verge of ineffectiveness, having approached and passed the ninety-day window that defined the limits of their utility. Stilwell seemed determined, however, to squeeze every last drop of blood out of this magnificent light infantry force—a force that he never seemed to understand or appreciate.

The costs of such misuse were ghastly. "Crucified" by Stilwell (according to historians Raymond Callahan and Louis Allen), Galahad eventually suffered 80 percent casualties. Most of these 2,400 casualties came from nonbattle sources. Nonetheless, many of the casualties could have been avoided through humane treatment of the men. Stilwell's staff even went so far as to roust still unrecovered soldiers of the 5307th out of their hospital beds in the rear to be flown in for the grist mill at Myitkyina.

Calvert's 3,000-man brigade numbered only 300 by the end of the battle for Mogaung. To avoid further commitment to combat, Calvert shut down his radios intentionally and marched his remnants to safety. Two other British commanders signaled that their battalions could no longer be counted on to obey an order to attack.[68] When the Morris Force was finally evacuated, its numbers had dwindled from 1,350 to a mere 50. Perhaps the most startling example of casualties belongs to the 111th. At the conclusion of his last directed tactical encounter, the British demanded that Master's brigade be evaluated by a team of doctors. Stilwell acceded to the request. Over 2,200 men were examined; only 118 were deemed fit for further service.[69] Incredibly, Stilwell then ordered this remnant to assume the defense of a Chinese artillery battery.

This last statistic starkly demonstrates that disease and exhaustion, not battle casualties, struck down most of the Chindits. Malaria, dysentery, diarrhea, undetermined fevers, naga sores: at least one of these ailments afflicted almost every Chindit. Not long into the campaign, another deadly disease made its appearance—scrub typhus. Galahad's casualties are strikingly illustrative of the imbalance between battle and disease losses. At Walawbum, Galahad lost 8 killed and 37 wounded; however, 179 other soldiers were evacuated—the victims of malaria (19), fevers (8), combat shock (10), injuries (33), and other illnesses (109). Galahad's losses at Nphum Ga were 57 killed in action, 302 wounded in action, and 379 incapacitated due to illness and exhaustion.[70] After the capture of the Myitkyina airfield, Galahad lost up to 100 men a day even though it was seeing much less action than before (see table 2).[71]

All the Chindit commanders watched with mounting horror and inner despair as their commands disintegrated before their eyes. Several of the brigades were ruined beyond help. Later, as Hunter recalled the loss of effectiveness and will to fight in the men, he stated one of the simplest and most important lessons of the campaign: "Sick men have no morale." In a similar vein, Calvert noted that he and his fatigued men began to avoid contact, to veer away from Japanese units, even when his unit was the superior force. Thus, the corollary to Hunter's dictum is Calvert's observation that exhausted men have no courage.

Table 2. Galahad Casualties

	Casualties	Percent	
		Actual	Preoperational Estimate
Battle Casualties:			
Battle deaths	93		
Nonbattle deaths	30		
Wounded in action	293*		
Missing in action	8		
SUBTOTAL	424	14	35
Disease Casualties:			
Amoebic dysentery	503		
Typhus fever	149		
Malaria	296**		
Psychoneurosis	72		
Miscellaneous fevers	950		
SUBTOTAL	1,970	66	50
TOTAL	2,394	80	85

*These are the official Adjutant General statistics. Many light battle casualties were treated on site and not evacuated, thus not reported. Complete statistics are not obtainable. The actual number of wounded at Nphum Ga exceeded the official total for the entire campaign.

**This is the number of malaria cases evacuated. Nearly every member of Galahad had malaria in a more or less severe form.

Operational Costs

One of the questions that most historians have raised in their discussions of the Chindits is whether or not the second expedition was worth its heavy costs. Beyond a doubt, Chindit II required a huge diversion of resources in two main areas, infantry and air support. This expenditure of resources can be measured easily. What is not so easy to clarify, however, is the degree of benefit that the Chindits produced. Could they have been put to better use as conventional formations fighting at Imphal and Kohima?

The diversity of opinion on the last question is extensive. Several distinguished historians believe that the Chindit operations had no impact on the Battle of Imphal-Kohima. Raymond Callahan has written that the Special Force never drew half as many Japanese into battle with them as they numbered themselves. He also criticized the Special Force as a misfit, a force too large to be a guerrilla force and too light to be a stronghold or assault force.[72] Field Marshal Slim has also discounted the influence of the Chindits, and even Brigadier Bidwell doubted whether or not the Chindits "paid their way," so to speak. On the other hand, Masters and Calvert have gone to some length to substantiate their claims that their brigades made significant contributions to the overall theater strategy. The testimony of the defeated Japanese commanders tends to support this view.[73]

Brigadier Calvert, in Chindit garb, at Mogaung

In a review of arguments concerning the Chindit operations, three conclusions have strong support. First, it is clear that Galahad's operations were indispensable to the advance of Stilwell's Chinese. In the Walawbum battle alone, Galahad's appearance in the rear of the 18th Division led Tanaka to fall back farther in one week than he had in the previous three months.[74] Without Galahad, Stilwell would never have made it to Myitkyina in 1944. There is no question that Galahad had more value in the enemy rear than it would have had attacking from the front. Moreover, had the First Chinese Army or Chiang's army been bolder and more aggressive in their advances, Stilwell could have achieved even more than he did and probably at less cost to Galahad.

Second, a Special Force composed of six brigades was undoubtedly a larger force than that required for its stated objectives. Two or three brigades, properly used, would have been sufficient to cut lines of communication to the 18th Division.[75] The 16th Brigade spent most of its time marching 450 miles to Indaw to attack a questionable objective that it ultimately did not take. It was then withdrawn. The 14th Brigade wandered fruitlessly from one area to another and saw little action, yet it still suffered significant casualties to disease and fatigue. The West African Brigade saw more action than the 14th, but it, too, produced questionable benefits overall. The 23d Brigade was never employed as a Chindit brigade; instead, it fought at Kohima in a conventional role. Only the 77th and the 111th Brigades carried their weight fully (though no criticism of the other brigades is intended). In retrospect, the 70th Indian Division should not have been broken up into Chindit brigades. It would have had a much greater impact on line as a division at Imphal.

40

Third, the Chindits could have been effective had they been used properly and directed at single goals commensurate with their capabilities. Slim's own self-criticism for misemploying the 3d and 14th Brigades supports this view. The concept of long-range penetration retains validity regardless of whether or not the Chindits implemented it perfectly in 1944. The question for historical analysts to resolve is not *whether* the Chindits were appropriately effective, but rather how they *could have been* employed optimally.

Conclusions

The Chindit War has great value to military historians and analysts because of the many conclusions that can be drawn from it regarding light infantry operations in an enemy's rear. Foremost among these is the observation that light infantry forces can be employed at the operational level of war. However, the caveat must be attached that effective light infantry operations must be coordinated directly with conventional operations by the main armies. Failing this, light infantry forces will meet an unhappy doom. It is also clear that conventional forces must be willing to take high risks on occasion, if they are to exploit fully the achievements of the light infantry. Stilwell failed to take such risks, particularly at Myitkyina.

The importance of air superiority to light infantry forces like the Chindits must also be considered. The Chindits depended utterly on their air lifeline, and the air lifeline depended utterly on air superiority. Had the Japanese been able to contest control of the airspace, it is unlikely that the Chindits would even have been committed. Beyond a doubt, air superiority enabled the Chindits to operate deeper, with more secrecy, and over a longer period of time than would have been possible otherwise. Mounting a similar operation today, in mid-intensity war, would require a similarly ingenious method of sustaining the force in the enemy's rear. Supply by air would be difficult to achieve.

Another significant conclusion of the Chindit War is that light infantry can perform both light infantry tasks (long-range penetration) and conventional tasks, but that the latter poses risks to light infantry and fails to take advantage of its special skills. Moreover, the mere existence of a large light infantry force in a theater increases the likelihood that it will be misused. Commanders are loath to leave a force uncommitted when there are so many objectives to be taken.

The experience of the Chindits shows that if light infantry is used conventionally, it must be augmented with heavy weapons. If blocks are established, the safety of the blocking force is enhanced if there are strong floater elements nearby to attack the enemy in its rear or flank.

Tactically, the Chindits demonstrated that a properly trained unit in the rear of an enemy can have an effect far out of proportion to the actual numbers of men involved. Furthermore, a small force can defeat a larger force if it achieves surprise and attacks the enemy where it least expects an attack. Chindit operations also show that enemy rear objectives must be sufficiently deep to guarantee against enemy reinforcement. Choosing objectives that are too shallow (for example, Blackpool) risks the engagement of light infantry by the enemy's main forces.

41

Four days' rations for a Chindit: biscuits, cheese, meat or "spam," sugar, salt, chocolate, tea, matches, powdered milk, cigarettes

Clearly, rear area operations require some knowledge of native and enemy languages. Chindit operations were materially enhanced by the Nisei speakers in their units and by the support received from the native populations, particularly by the Kachins. The tremendous Kachin support illustrates how valuable indigenous resources can be to light infantry, provided they are ready to take advantage of the resources. In short, light infantry forces must be trained to use whomever and whatever the environment offers.

The high value of the individual skills and special tactics of light infantry is borne out by the Chindit War. The Chindits proved themselves superior in jungle craft to the Japanese, because they had trained hard and adapted themselves to the environment. They modified their tactics to exploit the terrain better than the Japanese. Galahad operations exhibited a number of tactics that remain a model for light infantry: a swift approach march along an unguarded route; the retention of surprise; a hasty, accurate reconnaissance, followed by a bold attack against the enemy's weakness; and the employment of well-aimed, disciplined fires. The Chindits also demonstrated the power of the highly trained individual soldier and the necessity for his expertise in the basic skills of marksmanship, land navigation, stealth, scouting, and endurance. Chindit work, however, was young man's work; the old and infirm quickly became casualties. The experience of the Chindits also demonstrates that good infantry soldiers can be converted into good light infantrymen if given the proper training.

The Chindits were not supermen. Eventually, strain, stress, and fatigue affected them all. A limit exists as to how long a unit can be expected to remain effective in the enemy's rear. Evacuation of such men after ninety days seems to be right on the mark. Of course, under different circumstances, the period could be shorter or longer.

Finally, the Chindits demonstrated the psychological impact that a light infantry force can have against the enemy and on their own forces. The Chindits shocked the Japanese. The superior tactical mobility of the Chindits surprised them at first, but they were also stunned to meet Allied units that could stand, fight, and defeat them. Eventually, the predatory actions of Galahad and the Special Force led to an erosion of confidence among the Japanese troops.

Conversely, the Special Force boosted the morale and confidence of their own army. They demonstrated that the Japanese could be defeated, and they showed how it could be done. They showed that any good soldier could use the jungle to his own advantage. In short, they infused new life into the Fourteenth Army and fostered a winning spirit in it. Perhaps, the highest compliment paid to the Chindits came from Lord Mountbatten. Explaining by letter to Calvert that he intended to dissolve the Chindits, Mountbatten wrote, "It was the most distasteful job in my career to agree to your disbandment, but I only agreed because by that time the whole Army was Chindit-minded and therefore there was no need for a Special Force as such."[76] By this, he surely meant that the entire army had overcome its apprehension about the jungle and about the Japanese. Taking a lesson from the Chindits, the Fourteenth Army went on to defeat the Japanese Army in Burma in the Chindit style of boldness, aggressiveness, and confidence.

NOTES

Chapter 1

1. Stewart Alsop and Thomas Braden, *Sub Rosa: The O.S.S. and American Espionage* (New York: Reynal and Hitchcock, 1946). The exploits of the Kachin levies are explored vividly in this book.

2. Christopher Sykes, *Orde Wingate: A Biography* (New York: World Publishing Co., 1959), 432.

3. William Slim, 1st Viscount Slim, *Defeat Into Victory* (New York: David McKay Co., 1961), 135.

4. Ibid., 224.

5. The diversion of one column of the 111th Brigade to the Morris Force had not been planned. The column was diverted because of the delay in its deployment. The main body could not wait in the Chowringhee area for the column to assemble and join up, so it was sent east to join the Morris Force. This change in plan shows the flexibility of the Chindits—their ability to react prudently to unforeseen situations.

6. Wingate had announced widely that the Chindit brigades would be withdrawn after ninety days of operations. All the Chindits, including Galahad, accepted this limit as an article of faith. Events subsequently proved Wingate to be absolutely correct about the disastrous effects of leaving a Chindit unit in the field beyond this time limit.

7. Michael Calvert, *Prisoners of Hope* (London: Leo Cooper, 1971), 40.

8. Charles James Rolo, *Wingate's Raiders* (New York: Viking Press, 1944), 195.

9. Sykes, *Orde Wingate*, 372, 432. A higher percentage of older men was lost in both the training and combat phases.

10. Shelford Bidwell, *The Chindit War: Stilwell, Wingate, and the Campaign in Burma, 1944* (New York: Macmillan Publishing Co., 1979).

11. Ian Fellowes-Gordon, *The Magic War: The Battle for North Burma* (New York: Charles Scribner's Sons, 1971), 32.

12. Charles Newton Hunter, *Galahad* (San Antonio, TX: Naylor Co., 1963), 2. See also Charles F. Romanus and Riley Sunderland, *Stilwell's Command Problems*, United States Army in World War II (1956; reprint, Washington, DC: Office of the Chief of Military History, United States Army, 1970), 34.

13. Louis Allen, *Burma, the Longest War, 1941—1945* (London: J. M. Dent & Sons, 1984), 319.

14. David Halley, *With Wingate in Burma* (London: William Hodge and Co., 1946), 7.

15. Calvert, *Prisoners of Hope*, 17; and John Masters, *The Road Past Mandalay* (New York: Harper & Brothers, 1961), 193.

16. Rolo, *Wingate's Raiders*, 46.

17. Halley, *With Wingate in Burma*, 31.

18. U.S. War Department, General Staff, *Merrill's Marauders (February—May 1944)*, American Forces in Action Series (Washington, DC: Military Intelligence Division, U.S. War Department, 4 June 1945), 15, hereafter cited as *Merrill's Marauders*.

19. Masters, *Road Past Mandalay*, 188—89. Masters states, for example, that it only took fifteen minutes for a column to organize a night defensive position.

20. Hunter, *Galahad*, 7.

21. Ibid., 20.

22. Charlton Ogburn, *The Marauders* (New York: Harper & Brothers, 1956), 92.

23. *Merrill's Marauders*, 94—98.

24. Ogburn, *Marauders*, 229—31.

25. Calvert, *Prisoners of Hope*, 282—83.

26. Bidwell, *Chindit War*, 256.

27. Masters, *Road Past Mandalay*, 178—205; and Calvert, *Prisoners of Hope*, 100.

28. Calvert, *Prisoners of Hope*, 39.

29. Bidwell, *Chindit War*, 142.

30. Ibid.

31. *Merrill's Marauders*, 14.

32. Fellowes-Gordon, *Magic War*, 83; and *Merrill's Marauders*, 32.

33. Ogburn, *Marauders*, 141.

34. James Baggaley, *A Chindit Story* (London: Souvenir Press, 1954), 154.

35. *Merrill's Marauders*, 74.

36. Ibid., 69. Ogburn, *Marauders*, 197, reports making 100 river crossings in a 5-day, 70-mile march.

37. Ogburn, *Marauders*, 194—96.

38. Ogburn gives one of the best examples of this type of attack. Ibid., 168—71.

39. Calvert, *Prisoners of Hope*, 40.

40. Ogburn, *Marauders*, 113.

41. Hunter, *Galahad*, 13; and Fellowes-Gordon, *Magic War*, 66.

42. One notable exception to this practice occurred in early April when Stilwell's staff ordered Galahad out of its blocking positions near Shaduzup and Inkangahtawng into a defensive position at Nphum Ga. There, Galahad withstood an eleven-day siege by two Japanese battalions. Galahad suffered over 700 casualties during the siege.

43. *Merrill's Marauders*, 43.

44. Calvert, *Prisoners of Hope*, 142, 163.

45. Alsop and Braden, *Sub Rosa*, 198. According to this source, the Kachin Rangers killed 5,447 Japanese, captured 64 enemy soldiers, rescued 217 airmen shot down behind the lines, at the cost of 70 dead Kachin Rangers and 15 dead U.S. advisers.

46. Ibid., 193.

47. Ibid., 189.

48. *Merrill's Marauders*, 25.

49. Rolo, *Wingate's Raiders*, 63.

50. *Merrill's Marauders*, 26.

51. Ogburn, *Marauders*, 168—71; and *Merrill's Marauders*, 56.

52. Virtually every personal account of the campaign complains bitterly about the inadequate supply of food. Ogburn, *Marauders*, 153, describes the fantasies and the rituals associated with eating food. Occasionally, Chindit columns missed their resupply airdrops. When this happened, they often went without food until the next drop could be coordinated. A few men resorted to stealing food from their comrades; when discovered, they were severely punished.

53. *Merrill's Marauders*, 25. See also James H. Stone, *Crisis Fleeting: Original Reports on Military Medicine in India and Burma in the Second World War* (Washington, DC: Office of the Surgeon General, Department of the Army, 1969). This source vividly describes the medical history of the Chindits and Galahad.

54. Calvert attempted to instill an aggressive spirit in his men. In addition to the points cited in this paragraph, Calvert also insisted that the Chindits should be extra cautious only when they were stalking their prey. Otherwise, he advised, they should persist in attacking the enemy to maintain a spirit of boldness. He said that no soldier inadvertently meeting the enemy would be wrong in shooting him. Calvert, *Prisoners of Hope*, 129.

55. Terrence O'Brien, *Out of the Blue: A Pilot with the Chindits* (London: Collins, 1984). According to O'Brien, Morris frequently avoided contact with the Japanese and became distressed when he had no orders.

56. Rolo, *Wingate's Raiders*, 242.

57. For a discussion of Stilwell and Galahad, see Scott R. McMichael, "Common Man, Uncommon Leadership: Colonel Charles N. Hunter with Galahad in Burma," *Parameters* 16 (Summer 1986): 45—57.

58. Bidwell, *Chindit War*, 83.

59. Fellowes-Gordon, *Magic War*, 10.

60. Ogburn, *Marauders*, 219.

61. Hunter, *Galahad*, 86, 193—94.

62. These bitter feelings are evident in the works by Hunter, Ogburn, and Stone.

63. Slim, *Defeat into Victory*, 233. Masters agreed completely with this assessment. Masters, *Road Past Mandalay*, 280.

64. Hunter, *Galahad*, 71.

65. Romanus and Sunderland, *Stilwell's Command Problems*, 191.

66. Bidwell, *Chindit War*, 254; and Masters, *Road Past Mandalay*, 281.

67. Bidwell, *Chindit War*, 261.

68. Ibid., 19.

69. Masters, *Road Past Mandalay*, 275.

70. *Merrill's Marauders*, 45, 90.

71. Fellowes-Gordon, *Magic War*, 124; and Ogburn, *Marauders*, 252. One combat group under Captain Tom Senff lost thirty-two men in two days just marching twenty miles to join the main body at Myitkyina.

72. Raymond Callahan, *Burma, 1942—1945* (Newark: University of Delaware Press, 1978), 139.

73. Various sources show that the Japanese diverted up to a division of troops to deal with the Chindit threat. The testimony of the defeated Japanese commanders is cited in Calvert, *Prisoners of Hope*, 297—99; and Bidwell, *Chindit War*, 173.

74. Bidwell, *Chindit War*, 101, 284; and Fellowes-Gordon, *Magic War*, 87.

75. Masters, *Road Past Mandalay*, 280. Masters insists that Wingate should have employed the Chindits in a more concerted fashion, rather than frittering them away against disparate objectives (particularly the 16th, 14th, and 3d Brigades). He also states that the Special Force should have been used against the Japanese lines of communication (LOCs) to the central front, not to the north. Galahad was sufficient to interdict LOCs to the north.

76. Calvert, *Prisoners of Hope*, 15.

Chapter 1

Allen, Louis. *Burma, the Longest War, 1941—1945.* London: J. M. Dent & Sons, 1984.

Alsop, Stewart, and Thomas Braden. *Sub Rosa: The O.S.S. and American Espionage.* New York: Reynal and Hitchcock, 1946.

Asprey, Robert B. *War in the Shadows: The Guerrilla in History.* Volume 1. Garden City, NY: Doubleday and Co., 1975.

Baggaley, James. *A Chindit Story.* London: Souvenir Press, 1954.

Bidwell, Shelford. *The Chindit War: Stilwell, Wingate, and the Campaign in Burma, 1944.* New York: Macmillan Publishing Co., 1979.

Callahan, Raymond. *Burma, 1942—1945.* Newark: University of Delaware Press, 1978.

Calvert, Michael. *Prisoners of Hope.* London: Leo Cooper, 1971.

Cruickshank, Charles Greig. *SOE in the Far East.* New York: Oxford University Press, 1983.

Fellowes-Gordon, Ian. *The Magic War: The Battle for North Burma.* New York: Charles Scribner's Sons, 1971.

Fergusson, Bernard. *Beyond the Chindwin.* London: Collins, 1962.

_____. *The Wild Green Earth.* London: Collins, 1946.

George, John B. *Shots Fired in Anger.* Washington, DC: National Rifle Association of America, 1982.

Halley, David. *With Wingate in Burma.* London: William Hodge and Co., 1946.

Higgins, William J., Major, et al. "Imphal-Kohima: Encirclement." Student staff group battle analysis. U.S. Army Command and General Staff College, Fort Leavenworth, KS, 1984.

Hunter, Charles Newton. *Galahad.* San Antonio, TX: Naylor Co., 1963.

James, Richard Rhodes. *Chindit.* London: John Murray, 1980.

Masters, John. *The Road Past Mandalay.* New York: Harper and Brothers, 1961.

McMichael, Scott R. "Common Man, Uncommon Leadership: Colonel Charles N. Hunter with Galahad in Burma." *Parameters* 16 (Summer 1986):45—57.

O'Brien, Terrence. *Out of the Blue: A Pilot with the Chindits.* London: Collins, 1984.

Ogburn, Charlton. *The Marauders*. New York: Harper and Brothers, 1956.

Rolo, Charles James. *Wingate's Raiders*. New York: Viking Press, 1944.

Romanus, Charles F., and Riley Sunderland. *Stilwell's Command Problems*. United States Army in World War II. 1956. Reprint. Washington, DC: Office of the Chief of Military History, United States Army, 1970.

Slim, William Slim, 1st Viscount. *Defeat Into Victory*. New York: David McKay Co., 1961.

Stone, James H. *Crisis Fleeting: Original Reports on Military Medicine in India and Burma in the Second World War*. Washington, DC: Office of the Surgeon General, Department of the Army, 1969.

Sykes, Christopher. *Orde Wingate: A Biography*. New York: World Publishing Co., 1959.

U.S. Army Air Forces. 10th Air Force. "Development of Close Support Techniques in North Burma." 5 September 1944.

———. "Development of Joint Air-Ground Operations in North Burma." January 1945.

U.S. War Department. General Staff. *Merrill's Marauders (February—May 1944)*. American Forces in Action Series. Washington, DC: Military Intelligence Division, U.S. War Department, 4 June 1945.

Chapter 2

The Chinese Communist Forces in Korea

Introduction

In the autumn of 1950, forces of the United Nations (UN) Command, directed by General Douglas MacArthur, pushed confidently through the mountains of North Korea toward the Yalu River and the Manchurian border. Composed in almost equal parts of U.S. and Republic of Korea (ROK) divisions, with a sprinkling of other national forces, the UN Command advanced dreamily, even daring to forecast an end to the war by Christmas. Little did they know (despite adequate indicators) that a huge Chinese Army lay in wait, tensing for the right moment to pounce on the unsuspecting UN columns. At the appropriate instant, the Chinese Communist Forces (CCF) fell on the UN Command achieving strategic, operational, and tactical surprise, while attacking with such ferocity and shock that MacArthur's formations were pushed to the brink of disaster, reeling back under relentless Chinese pressure to a line well below Seoul (see map 8). That the Eighth Army, under General Matthew Ridgway, was able to recover and eventually restore the military situation in no sense detracts from this remarkable accomplishment by the CCF. This Chinese army—essentially a light infantry army—forms the subject of this chapter.

The Chinese Army's dependency on manpower (due to its shortage of military hardware) and the ruggedness of the Korean terrain determined the way the CCF was structured and employed during the Korean War. Compensating for its weakness in armaments and exploiting the possibilities of the rough Korean landscape, the Chinese developed a philosophy of "man over weapons" and organized a light infantry army to fight the war.

This army operated on a Korean peninsula whose physical and military characteristics have been described by S. L. A. Marshall as follows:

> There is no coastwise country in the world less suited than Korea to the movement of military forces in war, and there is none that offers so little comfort and reward to its conquerors.
>
> Almost the entire length of the country is mountainous, and the ridgeline heights are massive rock. At best, only small shrubs, stunted trees, and sparse grass maintain a foothold on the eroded slopes. There are no thick forests anywhere. The few hard-surface roads that run between the larger cities never have more than two lines of pavement, and this pavement is laid so thin that it cracks everywhere. Away from the main arteries of traffic there are only dirt tracks suitable for oxcarts and pedestrians. The few bridges that span the waterways are usually crudely built, and capable of handling one-way traffic

51

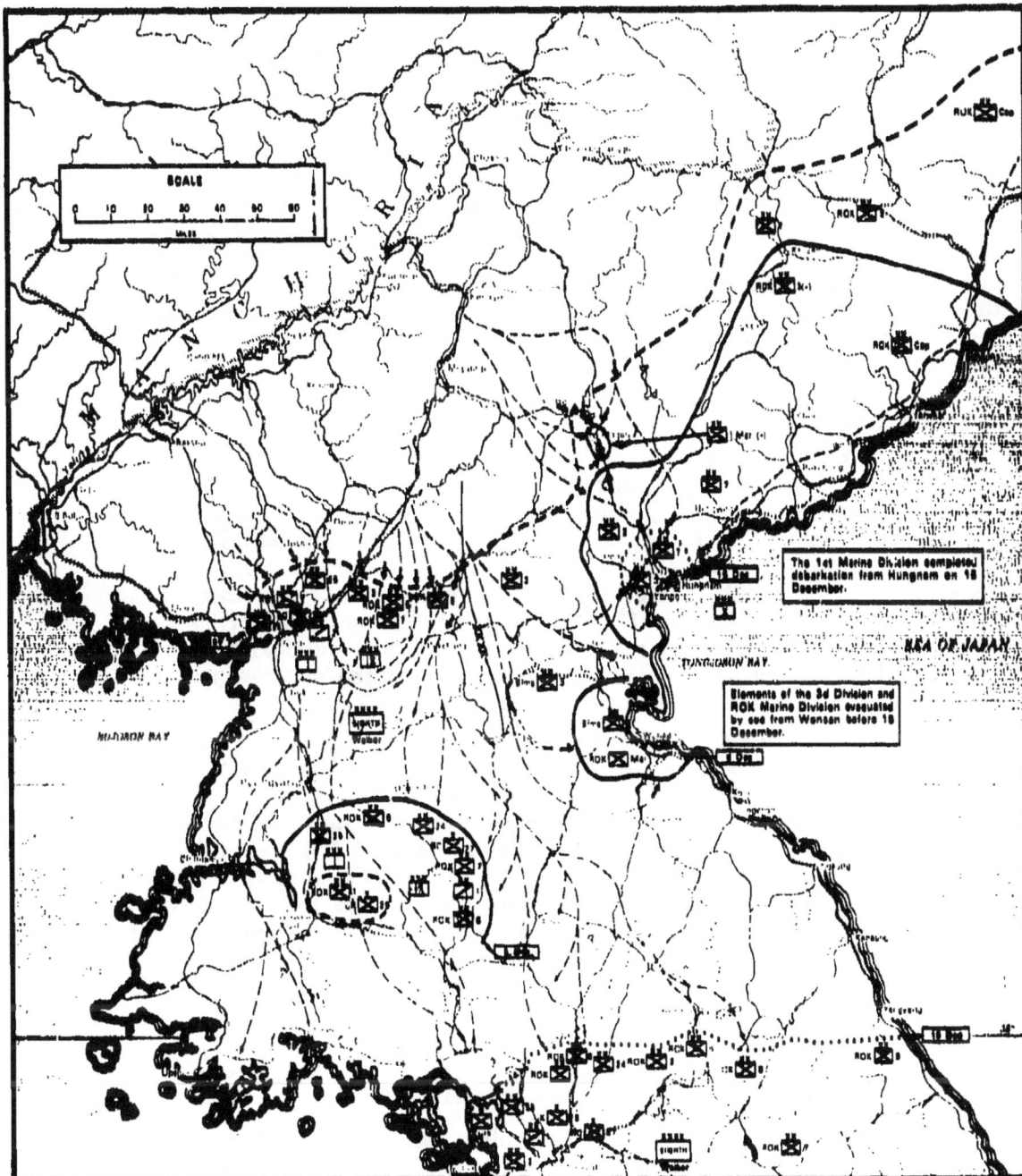

Map 8. Chinese Intervention in Korea, 24 November—15 December 1950

only. The usual river crossing is a ford. Most of the valley floors are so narrow that there is room only for a narrow path or a brook-size stream bed . . . ridges go on and on as far as the eye can see. All fighting in Korea is either uphill or downhill. Coping with the hills is more exhausting to fighting forces than meeting the fire of the enemy.[1]

Virtually every historian, analyst, or soldier who reflects on the Korean War points to the overwhelming influence of the terrain on military operations. Not only were combat actions constrained by the close terrain, logistical operations were also hindered, slowed, and severely interrupted by the poor transportation network and rugged escarpments. Under these conditions, the unique qualities and experiences of the CCF gave it, at least initially, a decided advantage over the less-spartan UN Command: "Without armor, with little artillery, unencumbered by complex communications, lightly equipped and carrying hand weapons only, the Chinese armies, which were inured to the extremes of weather and the scantiness of food, superbly disciplined and thoroughly trained, found choice opportunities here for maneuver and concealment."[2] From the beginning of the war, the CCF viewed the terrain as its ally, a combat multiplier against the heavier, road-bound UN forces. Indeed, the CCF viewed the terrain from an entirely different perspective than the UN Command.

But even had the terrain in Korea been amenable to operations by heavy forces, the Chinese High Command still would have retained its light organization—through necessity. Insubstantial military stocks, low levels of military aid from the USSR, and the absence of a military-industrial base in China to produce tanks, artillery, trucks, and aircraft dictated that the Chinese organize their only substantial military resource—manpower—into light infantry armies. Moreover, the Communist leadership, by virtue of its long experience as guerrilla warriors against the heavier armed Nationalist Chinese forces, had elaborated and refined its philosophy of "man over weapons" to compensate for CCF inferiority in weapons and materiel.[3]

Understanding this philosophy is central to understanding how and why the Chinese operated as they did in the Korean War. The expression "man over weapons" was not an empty slogan. The CCF's leadership and its soldiers firmly believed that by exploiting their superior human capabilities they would inevitably achieve success over the machine-burdened UN Command. This doctrine, they believed, had a certain moral strength to it, a spiritual power that guaranteed ascendancy on the battlefield. Furthermore, the Chinese were firmly convinced that when it came to soldiering—to the unyielding discipline and sacrifice required of men in combat—that the American and ROK soldiers were no match for them. Certainly, the early successes of the CCF reinforced its ideas that weapons did not count and that men did.

Organization and Equipment of the CCF

The organization of the CCF varied widely over time and from unit to unit. Figures 3 through 5 and table 3 present a composite picture of the CCF army, infantry division, infantry regiment, and infantry battalion.[4] (The CCF did not utilize corps headquarters. The army, essentially, was a corps.) These figures and table should be considered representative but not necessarily exact.

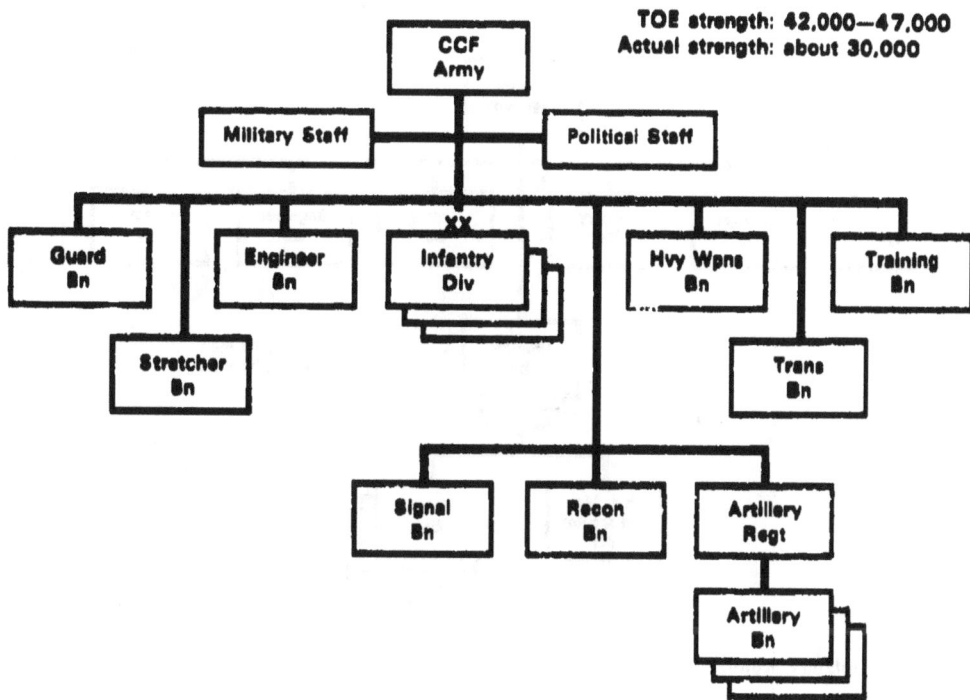

Figure 3. CCF army

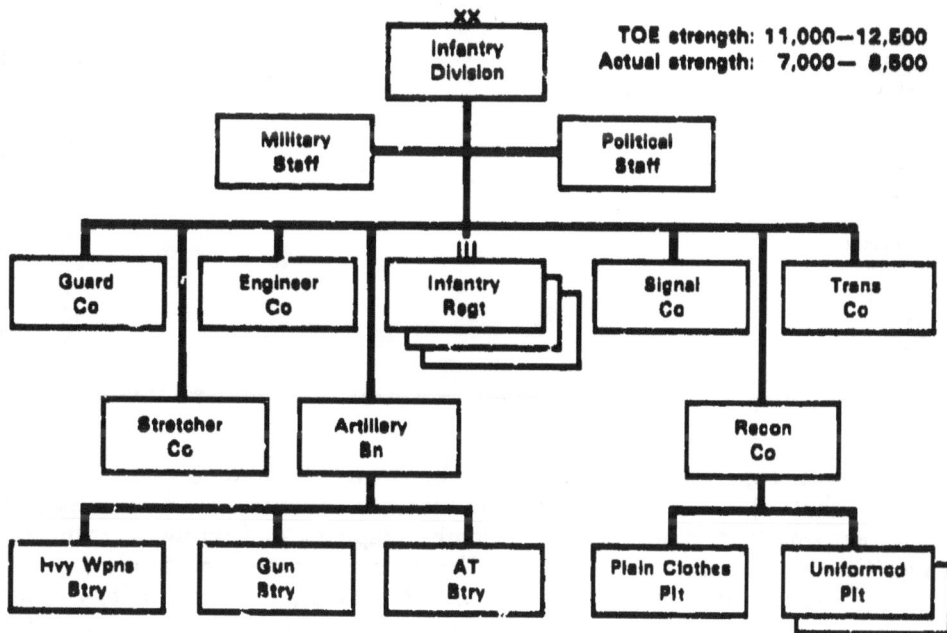

Figure 4. CCF infantry division

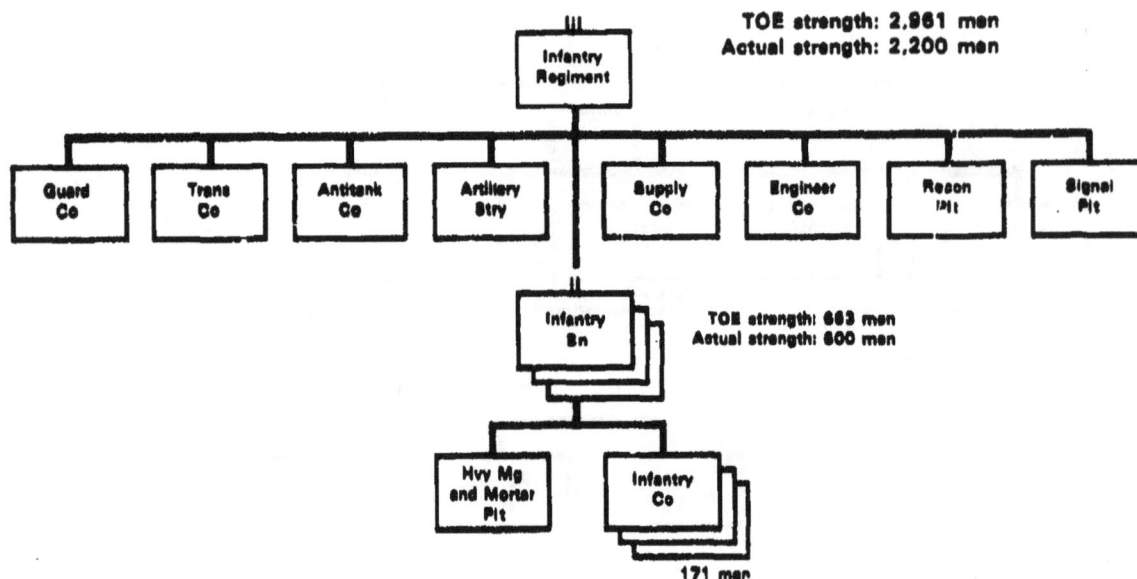

```
                                    ┌───┐
                                    │ III │
                                  ┌─┴───┴─┐        TOE strength: 2,961 men
                                  │Infantry│        Actual strength: 2,200 men
                                  │Regiment│
                                  └────────┘
```

Figure 5. CCF infantry regiment

Table 3. Equipment Totals per Regiment

Item	Number
Submachine guns	168
Light machine guns	94
Heavy machine guns	27
50-mm mortars	30
60-mm mortars	25
81- and 82-mm mortars	12
Antitank guns	6
75-mm guns	3
Bazookas	4
Trucks	20
Horses	405
Carts	40
Radios	5

The two most apparent observations to be drawn from inspection of these charts are that (1) every level of Chinese infantry organization from battalion to army suffered from a severe lack of combat support, and (2) motor transport in significant amounts did not exist. The meager equipment totals put the burden of the fighting on the infantrymen. Moreover, the lack of long-range systems ensured that CCF attacks would be conducted and decided at close range. Finally, the absence of transport guaranteed that operations would proceed no faster than the foot pace of CCF soldiers.

General Tactical Style of the CCF

In contrast to the wide variations that existed in CCF organization, CCF units generally exhibited a uniform tactical style. This common tactical style manifested itself in the use of surprise, deception and camouflage, movement, patrolling and reconnaissance, and individual skills.

Of these elements of tactical style, the Chinese attempted to employ surprise in almost every operation. And they frequently succeeded, in part, because they did things to achieve surprise that no western commander would dream of having his soldiers do.[5] Most of these measures fall into the category of deception and camouflage.

Deception and Camouflage

Chinese secret preparations for their initial assault on the Eighth Army in November 1950 typified their techniques of deception and camouflage. As part of this operation, up to 300,000 soldiers, undetected by U.S. intelligence, moved into secret assembly areas. Moving almost exclusively at night and hiding during daylight, whole rifle platoons were packed sometimes into single Korean huts, where they remained until dark. If shelter was not available, the soldiers lay huddled and motionless in ditches, gullies, and draws, covering themselves with straw, mud, and other materials. Sometimes forced to move by day in later operations, the Chinese picketed the hills along the route with observers. If a UN aircraft was spotted, the observer fired a shot, signaling troops within hearing to take cover. At times, troops on the march during the day carried straw mats on their backs. When lying down on the ground in orderly rows, the troop formation gave the appearance of a recently cut field to air observers. Some troops dressed in white, like the Koreans, and moved openly on the roads.[6] Others pretended to be ROK soldiers.

Vehicles were camouflaged by day or hidden in tunnels, under bridges, or in dugouts. The Chinese also parked operational vehicles among vehicles that had been destroyed or disabled in previous air strikes, or they left them in awkward positions in ditches to appear disabled. Such measures often fooled pilots during subsequent air strikes. In the same deceptive manner, the Chinese camouflaged destroyed vehicles to cause the UN air forces to waste their ordnance on what they thought were operational vehicles.[7]

Avoiding detection and choosing routes across difficult terrain and away from roads, the CCF infantry gathered its strength to attack. Prior to intervention, it initiated small combat actions to occupy the enemy's attention and to reinforce the myth that only weakness lay before the UN forces—a ploy described as "the pretense of picking around with a finger to cover the raising of a mailed fist."[8] Their actions were extremely effective; only deep patrolling by strong combat patrols would have detected them. The terrain enhanced the deception effort.

In rear areas or in assembly areas, CCF troops dug two-man or squad-size foxholes on reverse slopes. The soldiers carefully distributed the excavated soil, covering it with branches and straw and replanting the turf. These foxholes were cunningly selected and camouflaged to blend in with the terrain; they were virtually undetectable except from close range.[9] The Chinese also selected bivouac areas in or near burned-out villages, often taking up residence

CFF Method of camouflaging vehicles

among the rubble. The CCF avoided high ground, which characteristically attracted air strikes.[10] In addition, the Chinese used natural materials almost exclusively for camouflage and concealment, employing them with great imagination. UN patrols sometimes approached within killing range of the CCF, but the Chinese often did not fire, choosing to remain hidden.

Setting fires to produce smoke to cover movement was another CCF technique. During May 1951, they burned debris and vegetation for four days, obscuring the entire eastern part of the Eighth Army's lines to conceal the shifting of Chinese forces for a major offensive.[11]

The Chinese also intensified their night activities, which protected them from the prying eyes of UN aircraft. The CCF's willingness to act at night, coupled with the tendency of the UN Command to remain in a static defense, gave the Chinese a great advantage.

Patrolling and Intelligence

Heavy Chinese reliance on thorough reconnaissance and intelligence gathering prior to every operation also enhanced their war efforts. These activities took many forms but were usually the function of the reconnaissance companies organic to each infantry division. Regular infantry squads were used to reinforce or support these reconnaissance units, but they were rarely sent out on their own.

Reconnaissance companies were composed of two uniformed platoons and one plainclothes platoon. The uniformed platoons, composed of two officers and three squads of ten men, performed missions of reconnaissance patrolling. Most of the patrols sent out were of squad size and were armed with rifles and submachine guns. Generally, they first conducted reconnaissance to within a few kilometers of UN lines to obtain local guides, learn routes, become familiar with the terrain, and prepare to guide parent units into forward assembly areas. These patrols avoided contact.

As the date of an attack drew nearer, reconnaissance patrols approached right up to outposts and the main line of resistance. There, they tried to draw fire upon themselves to identify fighting positions, locate boundaries and flanks, and discover weak spots. In scouting out a particular objective, patrols normally took the shortest route to the site, usually moving in single file with no flank or rear guards. Once near the objective, the patrol separated, each member accomplishing his specific task, with all members reassembling at a predesignated point and returning to camp, often by the same route as they had come.

Patrols usually moved along low ground or below the crests of ridges and mountain sides. Patrol formations varied with the terrain but remained prescribed to a few specific variations. Patrol leaders often took the lead, but when they moved through dangerous ground, they used one- to three-man points.[12]

Chinese patrols were always vulnerable to ambushes due to their rigid adherence to standing operating procedures (SOPs), their failure to use flank or rear guards, and their habit of using the same routes to and from objectives. Nonetheless, Chinese patrols continued to follow these methods even after falling victim to UN ambushes. In spite of their limitations, however, Chinese patrol methods worked well. Most accounts of combat actions in Korea credit the CCF with an uncanny ability to locate the weakest sector, flank, or boundary between U.S. units.

Whenever uniformed patrols went behind enemy lines, they were reinforced with regular infantry for support. Most of the time, however, deep patrolling was undertaken by plainclothes platoons. The most common tactic of these plainclothes units was to infiltrate the UN lines through normal traffic or the refugee stream. Weapons were concealed on the person of individual Chinese soldiers or in accompanying carts. Having cleared the UN security screen, these forces then scouted out UN boundaries, positions, and rear areas before exfiltrating.[13]

The Chinese also placed a high value on intelligence obtained from local villagers. As mentioned above, they employed native guides, both voluntarily and involuntarily, to assist patrols and to guide units into positions. In addition, capture teams were sometimes sent out to seize UN soldiers for interrogation. Through the use of these methods—reconnaissance patrols, plainclothes unit infiltration, and local sources of intelligence—the CCF was always well informed in tactical intelligence prior to any attack.

Movement, Maneuver, and Infiltration

The CCF rarely used roads for movement of troops into assembly areas or attack positions. Instead, relying on the information provided by reconnaissance patrols, they moved cross-country. The CCF's high level of physical fitness and its recognition that the terrain offered concealment and protection permitted it to cover large distances on foot. Conducting long approach marches by night and hiding and resting by day, the Chinese continually surprised U.S.-UN observers with their ability to carry huge burdens on foot, as they rapidly covered long distances in the most hostile weather.

The Chinese did not walk or hike; they moved at a steady run that they could keep up for hours. Lacking radios, they moved in single files and column formations, with scouts out front to maintain command and control.[14] Occasionally, leaders were mounted to control movement better. S. L. A. Marshall provides a vivid description of this kind of movement, reporting the experience of Private First Class Louis Giudici from his perimeter guard post:

> At first there were just small groups of men, moving about six in a bunch with a 10-yard interval between them. They moved at double time, and though there were five of these small groups, none seemed to be carrying small arms. He had sighted them first at about 250 feet. The first were just drawing abreast of him when he reached for the trigger with the intention of opening fire.
>
> But things had changed, and he stayed his hand. A whole column of enemy infantry was now pouring into the creek bed, right on the heels of the reconnaissance groups. They seemed to be very large men, perhaps because the conspicuous white bandoliers which crossed their breasts and the overcoats which almost touched the ground increased their bulk. They carried rifles and tommy guns at the port as they, too, moved down the creek bed at a run. The column was four abreast. With every company or so rode a man on horseback, who shouted orders at the others as he moved along.
>
> For seventeen minutes this solid column moved at a run past this nineteen-year-old gunner, its closest files within 35 yards of his weapon. The time interval shows that at least one Chinese regiment raced by. They did not see him, and he felt that if he fired, it would mean the destruction of the company.[15]

Such units were well trained in immediate-action drills. Taking unexpected fire while still in march formation, they quickly took cover, then formed assault elements to eliminate the threat before reforming into column and moving on.

Rivers and streams were no barriers to the Chinese. They used existing bridges and fords where possible and, at other times, improvised bridges and rafts, which they could hide or dismantle by day. Typically, the CCF chose the physically easiest crossing sites, not the tactically best-positioned ones. Despite the cold, the hardy soldiers often waded and swam rivers at night in multiple columns. Speed and security were their main concerns.[16]

During the first year of the war, before the UN command was able to tie in its flanks from coast to coast, movements like those described above permitted the CCF to infiltrate between UN units. Many times, these infiltrations went undetected and unreported until the CCF actually attacked the flanks or rear of UN positions. At the close of the first Chinese offensive, an entire North Korean division infiltrated into the rear of the right flank of the Eighth Army, where it relied on the countryside for food and clothing and obtained arms and ammunition by raids on UN stocks. The U.S. Marine 1st Division and ROK security forces finally turned it around. Even so, the enemy division was able to maintain its coherence, break into small groups, and exfiltrate.[17]

Most infiltrations, however, were conducted by small units. One historian, in fact, has described the Chinese conduct of the war as an "endless succession of platoon infiltrations."[18] These small-unit infiltrations followed the patterns already described: thorough reconnaissance, adept use of terrain, identification of enemy unit boundaries and weak points along extended perimeters, noiseless movement, avoidance of UN patrols and outposts, and strict fire discipline. All of these tasks were accomplished at night. Successful infiltration depended on the abilities of the individual soldier. That the CCF was so adept at infiltration is a tribute to the unbelievable stealth of its soldiers. Indeed, the official U.S. Marine Corps history credits Chinese soldiers with being able to infiltrate at night better than any other soldiers on earth.[19] Innumerable accounts recall how Red soldiers were able to creep noiselessly within yards of UN positions and then rise up to attack violently.

Individual Soldier Skills

Stealth and infiltration were not the only skills of Chinese light infantrymen. In the absence of significant combat support, CCF tactical success required that Chinese soldiers possess many highly developed skills and attributes. Two of these attributes, physical conditioning and stamina, gave Chinese soldiers the strength to conquer the terrain and to keep pace with the more road-mobile UN forces. In fact, the CCF was far more tactically mobile than the UN command. In addition, the men of the CCF had the endurance to survive in the harsh, open climate, often on short rations. The fierceness and tenacity of Chinese soldiers gave their units a shock value that, when coupled with the surprise they achieved through stealth, often overwhelmed the superior firepower of the UN defenders.

Chinese infantrymen also knew how to use every fold in the terrain for cover and concealment. They seldom got lost or disoriented in the darkness and showed an unerring appreciation for the advantages of ground in the choice of their routes, placement of their machine guns and mortars, and in their selection of fighting positions.

CCF soldiers also enjoyed, at least initially, a mental advantage over their U.S. and ROK counterparts. Confident of their own superiority, scornful of American problems with the terrain, darkness, and weather, and convinced of their eventual victory, Chinese infantrymen had a psychological edge over their opponents. Moreover, their self-reliance enabled them to fight on against unfavorable odds and taught them not to depend on their own unreliable lines of supply. Indeed, they often sustained and equipped themselves with UN rations, arms, ammunition, and materiel.

The Chinese infantrymen of the Korean War were formidable opponents. Hardy, resilient, and tough, they earned the respect of their foes and carved themselves a niche in military history. Their skill and determination as light infantry soldiers were central to the tactical successes of the CCF.

The Attack

Prior to an attack, the CCF always performed thorough reconnaissance. Then, it accomplished its approach marches to forward assembly areas by night, in fast column formations. As the CCF closed with the enemy, separate combat groups split off to their sectors for the attack. Generally, combat groups entered their final assembly areas, ten kilometers or so from the UN lines, the night before the attack.

Surrendering Chinese soldier

With amazing regularity, CCF units from battalion to army level attacked while employing the following method. At dusk on the day of an attack, battalions moved out from their assembly areas and halted in sheltered positions one to two kilometers short of the UN lines. There, units took a short rest and perhaps ate a meal. Company commanders then received their orders and took charge of their units for the attack.

61

Each company subsequently moved out according to a detailed SOP. One company moved up to fix the enemy, while the other two companies attempted to envelop the flanks of the enemy position. Sometimes, only one company attempted envelopment, while the third company was held in reserve pending development of the situation. Timing seemed to be relatively fixed. Control lines were established between the rest position and the line of departure, which was about 200 meters from the UN lines (see figure 6).[20]

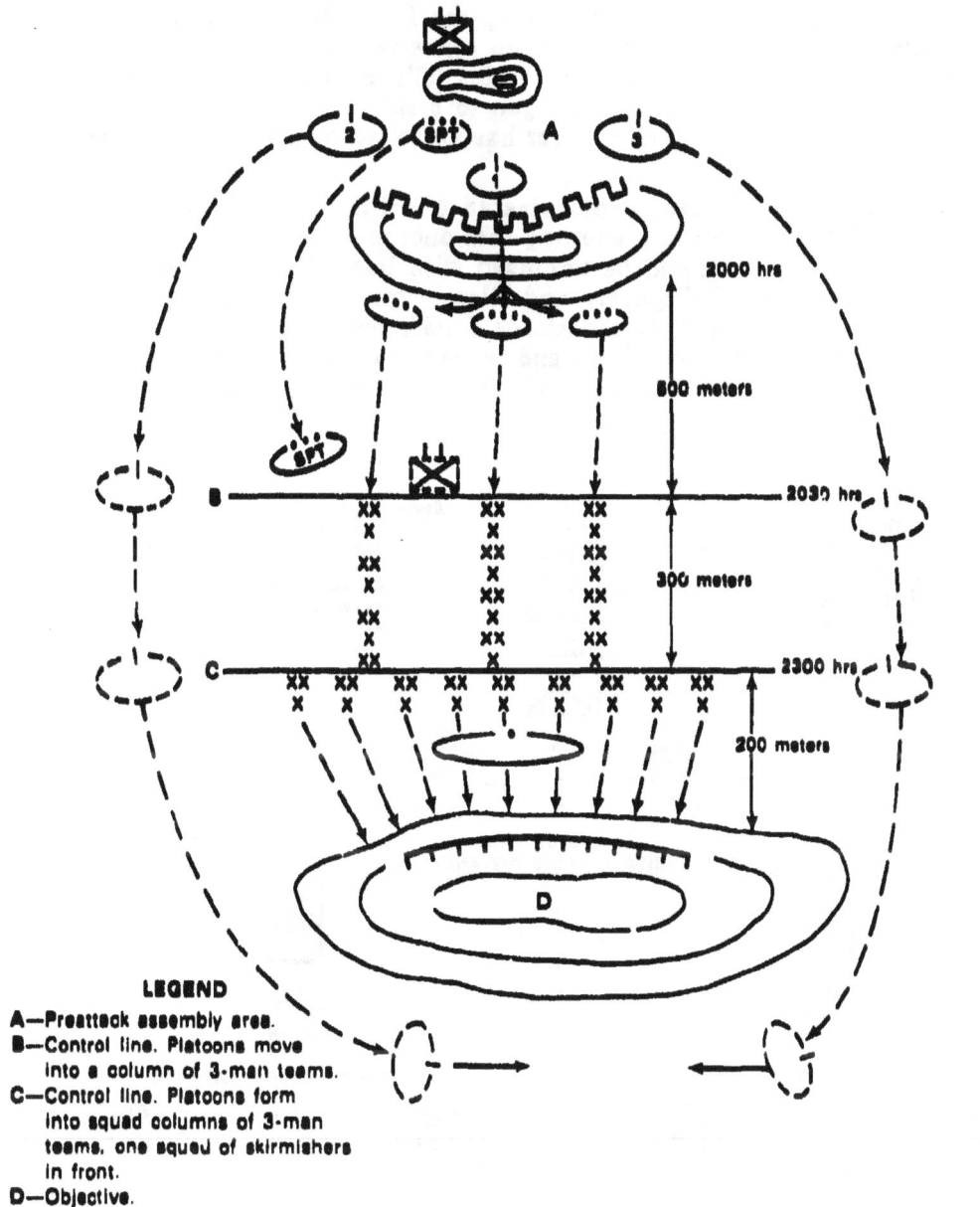

LEGEND

A—Preattack assembly area.
B—Control line. Platoons move into a column of 3-man teams.
C—Control line. Platoons form into squad columns of 3-man teams, one squad of skirmishers in front.
D—Objective.

Source: U S Army, IX Corps (Korea), "Enemy Tactics, Techniques and Doctrine," 14 September 1951, 17a

Figure 6. CCF battalion attack

The actual attack was normally launched between 2300 and 0100. First, lead elements in the attack tried to approach as close as possible to the enemy foxholes, with a thin skirmish line of scouts in front making the first contact. Many combat reports indicate that the CCF often succeeded in approaching to within 15 to 150 yards of the UN lines before being detected. On occasion, the first signs of an attack were exploding hand grenades lobbed by the Chinese from only yards away.

The Chinese company fixing the UN forces then maintained its pressure until the flanking companies began to roll up the flank. Meanwhile, soldiers practiced excellent fire discipline in order to conserve their ammunition for critical moments. Personnel also advanced during lulls in firing and took cover when necessary. As the UN resistance began to break, all the CCF elements then pressed forward in what one observer has described as an "assembly on the objective."[21]

In the attack, the Chinese demonstrated a willingness to take high casualties to maintain momentum, knowing that once one objective was taken, the rest of the enemy line could be unhinged with less effort. After seizing one enemy position, the CCF quickly and silently moved against other UN positions on the left or right. If the attack failed, however, Chinese units would break off the attack before dawn and retire to secure positions. (Figures 7 and 8 show how this type of attack might be conducted by a CCF army of three divisions.)[22]

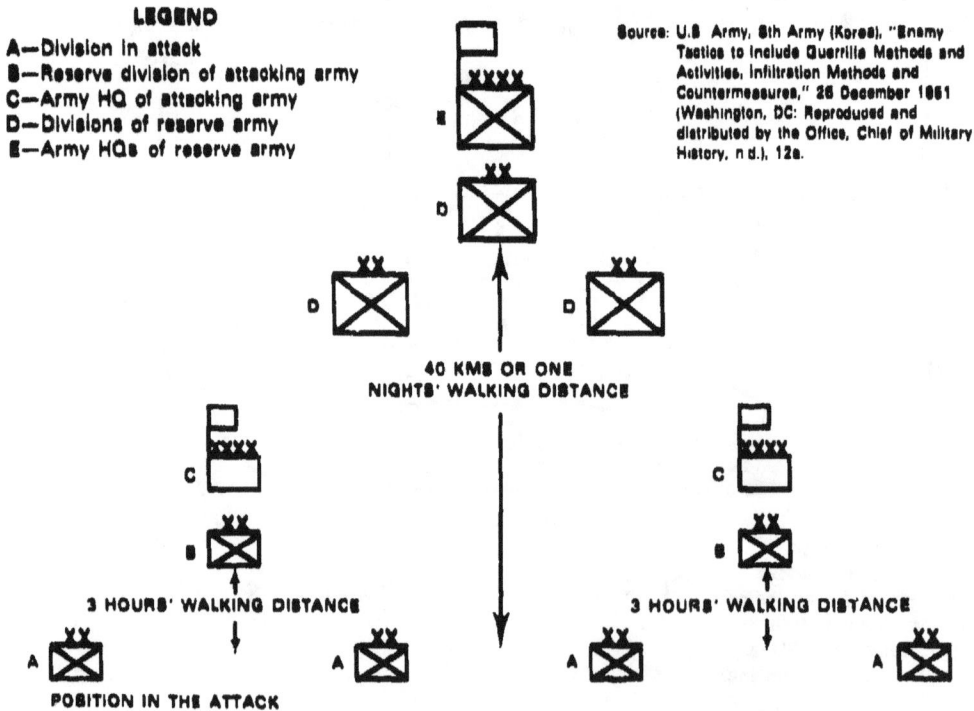

LEGEND
A—Division in attack
B—Reserve division of attacking army
C—Army HQ of attacking army
D—Divisions of reserve army
E—Army HQs of reserve army

Source: U.S. Army, 8th Army (Korea), "Enemy Tactics to Include Guerrilla Methods and Activities, Infiltration Methods and Countermeasures," 26 December 1951 (Washington, DC: Reproduced and distributed by the Office, Chief of Military History, n.d.), 12a.

40 KMS OR ONE NIGHTS' WALKING DISTANCE

3 HOURS' WALKING DISTANCE

3 HOURS' WALKING DISTANCE

POSITION IN THE ATTACK

Figure 7. CCF army approach march

(A)
PRIOR TO ATTACK

(B)
DEVELOPMENT OF ATTACK

LEGEND

EN—Enemy forces
1—Defensive CCF Div
2—Flanking CCF Div
3—Reserve CCF Div
——————Routes taken by divs

Area indicates fanning
out of defensive div

Probing attack

Source. U.S. Army, 8th Army (Korea),
' Enemy Tactics to Include
Guerrilla Methods and Activities,
Infiltration Methods and
Countermeasures," 26 December
1951 (Washington, DC:
Reproduced and distributed by
the Office, Chief of Military
History, n.d.) 48a.

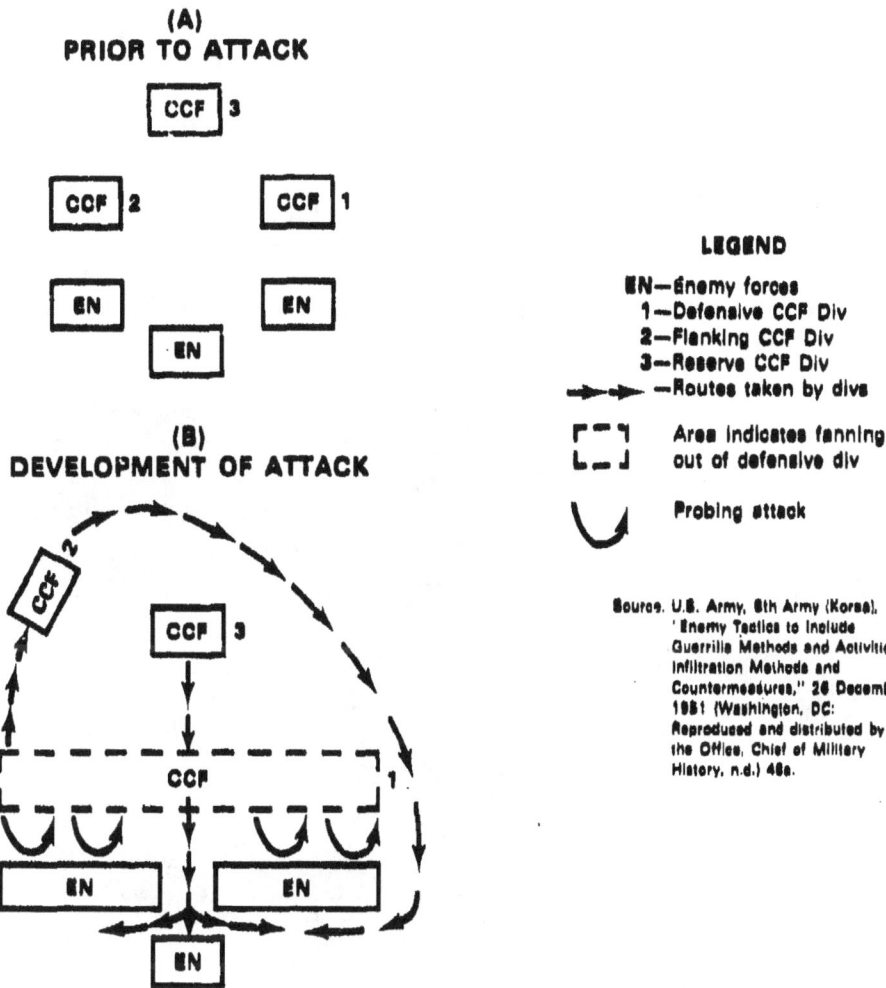

Figure 8. CCF army attack

Once committed, CCF units were permitted little flexibility. The same method of attack was used each time, and a battalion had to pursue the attack until its last rounds were fired or until it succeeded. Company commanders did not have the latitude to change the plan or call off the attack without permission.

Apparently, the main reason for this tactical rigidity in the attack was the problem of communications. Generally, radios were unavailable below regimental level, though wire was laid to battalion and sometimes company level in static situations. In the attack, however, communications at battalion level were based on runners and signals. (Sometimes, battalion and company commanders enjoyed the use of captured U.S. walkie-talkies.) Under these conditions, unyielding adherence to SOPs was the solution selected by the Chinese to guarantee command and control. Furthermore, to enhance command and control, the CCF preferred to attack on nights when bright moonlight could be expected.

64

During the attack, for a variety of purposes, the Chinese used bugles, whistles, flutes, and shepherd horns. To facilitate command and control, different tunes or notes meant different things: to advance, to increase fire, to cease fire, and so forth. The Chinese also used these signals to simulate to the enemy a more extended deployment than they had actually accomplished. The CCF also used the instruments to create an atmosphere of fear and terror before or during the attack, showing ingenuity in its methods. To obtain this effect, the CCF blew taps on the bugle, played haunting, eerie tunes on the flute, and created a cacophony of sound during the actual attack. Until American troops grew used to this tactic, they reacted with true alarm, especially at night.

An example of a Chinese artillery piece of ancient vintage used in Korea

Because of a continuing shortage of artillery and mortar rounds, the CCF did not rely on heavy bombardments prior to an attack. If artillery was used at all, it was used with discrimination because of the danger of its being detected and destroyed by UN artillery. Chinese artillery also was seldom massed, owing in part to a lack of training among the CCF artillery troops.[23] Mortars were employed more commonly and quite effectively. When artillery or mortar fires were used in quantity, Chinese troops followed closely behind the barrage. Nevertheless, it is important to note that Chinese indirect fires paled in comparison to those employed by the UN. When the CCF experienced a tactical stalemate, however, it began to rely more on artillery as a battlefield killer.[24] Consequently, in the last two years of the war, the Chinese made more extensive use of massed artillery fires to defend against UN attacks. The Chinese seldom used tanks to augment their firepower; they simply did not have sufficient numbers of them in their inventory, except for occasional use as mobile artillery.

At the soldier level, the preferred weapons were hand grenades and sub-machine guns. Each Chinese infantryman carried four to five grenades apiece. Even though 25 to 30 percent of them were duds and the Chinese were weak armed and could not throw them far, they used them time after time, believing U.S. soldiers were afraid of grenades.[25]

Disaster at Unsan

The Battle of Unsan, analyzed below, illustrates how the offensive tactics and techniques of the CCF worked in actual operations. In October 1950, the U.S. Eighth Army continued its steady advance into North Korea toward the Yalu River. The farther the army advanced, the thinner its lines stretched, the more extended its lines of communication became, and the greater the gaps between units grew. While the Chinese had not yet intervened in force, they monitored the progress of the UN Command with the same sharp-eyed interest that a wolf displays while watching a rabbit playing too far from its hole. On 30 and 31 October, the 8th Cavalry Regiment, leading the forward movement of the 1st Cavalry Division, occupied defensive positions on the high ground north and west of the town of Unsan (see map 9). Sent to bolster the 15th ROK Regiment, which was under heavy enemy attack, the members of the 8th Cavalry noticed the unusual presence of large smoke clouds hanging in the area. Unknown to them, the Chinese were setting forest fires to cover their movements. In fact, on occupying Unsan, the 8th Cavalry had walked unknowingly into a vicious trap. Here, the U.S. Army in Korea would experi-ence one of its first tastes of the Chinese style of war.

Unbeknown to the 8th Cavalry, the Thirty-Ninth Army of the CCF had infiltrated its 115th and 116th Divisions into the Unsan area. Moving with their characteristic stealth and discipline, the Chinese had thoroughly investi-gated the area and identified key routes and terrain. Although they had not fully pinpointed the American positions, the Chinese prepared to attack on the night of 1 November. Preparatory to this attack, five companies from the Thirty-Ninth Army established a strong blocking position across the road leading into Unsan, just west of the Turtle Head Bend of the Kuryong River—the obvious route to be taken by U.S. reinforcements if sent.

Meanwhile, on the afternoon of 31 October, the 8th Cavalry settled into its defensive positions. Across the Samtan River, the ROK 15th Regiment was slowly disintegrating in the face of the strong Chinese pressure. The sounds of this battle were clear; the men of the 8th Cavalry, particularly the 1st Battalion on the right flank, followed the progress of the combat with growing alarm.

As shown on map 9, neither flank of the 1st Battalion was connected to the units on the left and right. The bridge over the Samtan River was held by a platoon of tanks. Below the bridge, the right flank and rear of the battalion lay exposed, protected only as long as the ROK 15th Regiment held its positions.

At 1700, the battle spilled over into the 1st Battalion sector. At 1930, the enemy intensified its attacks, driving the right flank company in 400 yards. About 2100, the Chinese found the gap between the 1st and 2d Battalions and began infiltrating behind the Americans into Unsan. By 2200, the

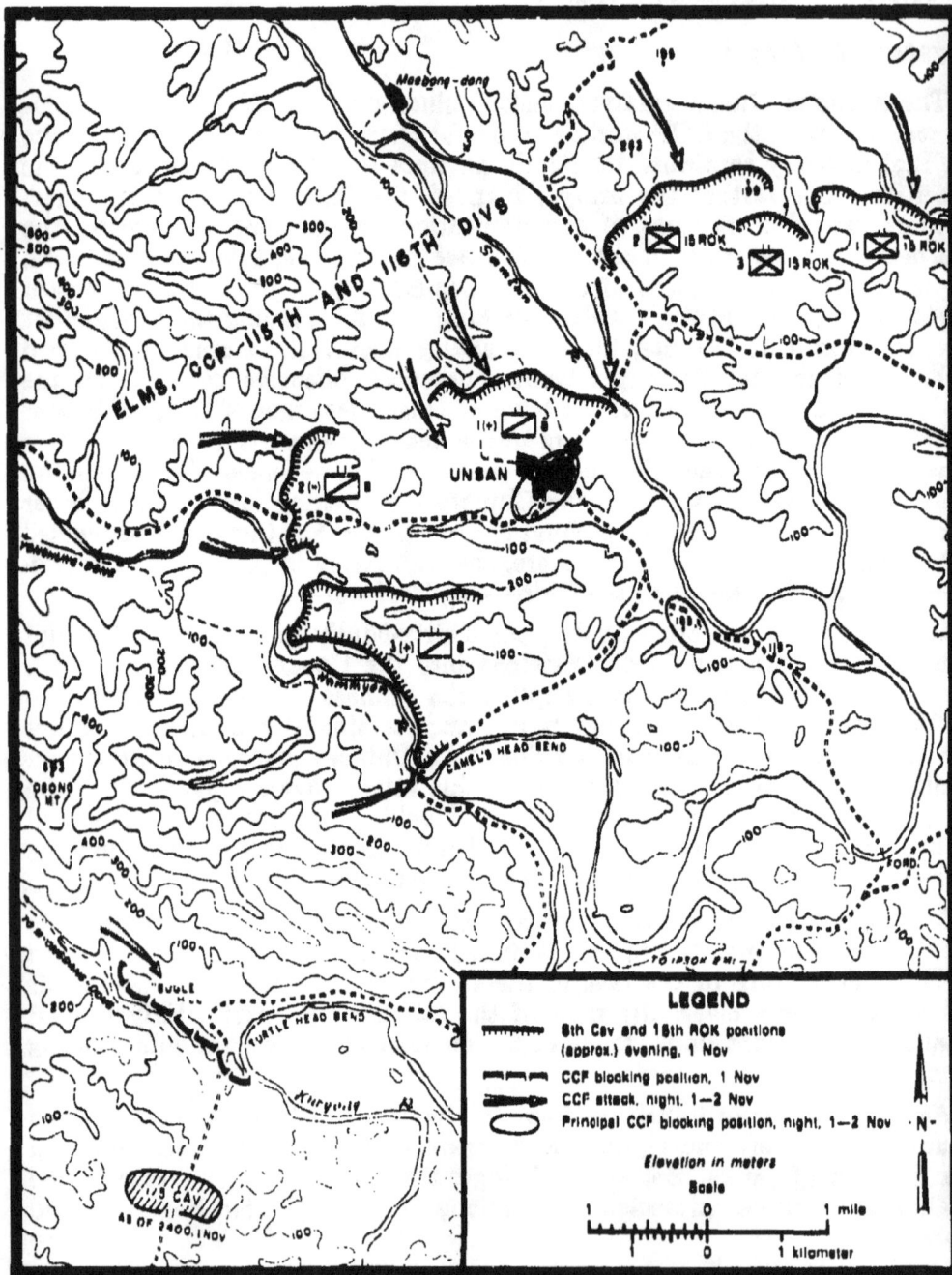

Source: Appleman. *South to the Naktong, North to the Yalu.* 692.

Map 9. The Unsan engagement, 1—2 November 1950

battalion commander, Major John Millikin Jr., realized that the Chinese held the right bank of the Samtan River. Recognizing that his position was tenuous, Major Millikin ordered his trains to withdraw through Unsan to the road fork about one and one-half miles south of the town, thence southeastward across the ford over the Kuryong River and into Ipsok about seven miles away.

After dark, the 2d Battalion also came under heavy attack. By 2300, the Chinese had penetrated both battalions and driven them from their primary positions. In the process, the battalions consumed most of their ammunition. The 3d Battalion was not yet threatened.

About this time, the 1st Cavalry Division ordered the regiment to withdraw. The regimental commander, however, decided to have the 1st Battalion hold Unsan while the 2d Battalion withdrew. Then, the 1st Battalion would withdraw. The 3d Battalion was ordered to keep the road fork below Unsan open until the 1st and 2d Battalions had cleared, then it was to withdraw also. The plan was a good one, but it came too late.

When Major Millikin moved to Unsan around midnight to direct the withdrawal of his battalion, he found the town occupied by Chinese. Consequently, he ordered his companies to bypass the town to the east and wait at the road fork below the town. Arriving there, Millikin discovered that his waiting elements were beginning to take small-arms fire from the Chinese in the area. Even worse, a small artillery convoy moving south of the road junction ran into a Chinese unit establishing a roadblock in the vicinity of Hill 135. Radio reports a few minutes later indicated that the Chinese also blocked the ford over the Kuryong. From this point on, no convoys were able to get out of the area. The road fork itself fell to a Chinese attack a short time later. The withdrawal of the 1st and 2d Battalions ended in a desperate attempt at escape and evasion by small groups over the next two to three days.

The 3d Battalion of the 8th Cavalry was the last to be hit by the CCF. Despite its awareness of the collapse of the 1st and 2d Battalions, this unit was also taken by surprise. About 0300, a company of Chinese crossed the bridge over the Nammyon River on the southern flank of the battalion position. The two squads guarding the bridge let them pass, thinking they were ROKs. Suddenly, the leader of the enemy column blew his bugle Within seconds, the company launched an assault against the battalion command post, and other waiting enemy forces attacked across the river.

From the start, the CCF had the upper hand in this fight, catching some of the Americans sleeping while they waited for the order to evacuate. The Chinese penetrated the U.S. lines immediately, and the attack dissolved into isolated but fierce actions against U.S. strongpoints.

At daylight, the strength of the 3d Battalion stood at 6 officers and 200 men. with 150 wounded. Out of range of supporting artillery, the battalion held . , during the day. Supporting air strikes did not help much because of the proximity of the Chinese and the continuing smoky haze.

The 1st Cavalry Division sent a relief column to extract the 3d Battalion, but the Chinese had correctly anticipated this maneuver. Two battalions of the 5th Cavalry Regiment could not break through the defensive block at

Turtle Head Bend. A battalion from the 7th Cavalry Regiment tried to go around the block, but it never entered the fight. At 1500 on 2 November, Major General Gay, Commanding General, 1st Cavalry Division, reluctantly ordered his command to withdraw, leaving the 3d Battalion to its fate. The remnants of the 3d Battalion fought valiantly through 4 November. Heavily mortared and reduced by repeated infantry attacks, the battalion then attempted its own escape on foot. Only about 200 men survived to rejoin their regiment. Total U.S. losses during the battle numbered approximately 600 men. The regiment also lost twelve 105-mm howitzers, nine tanks, and one tank recovery vehicle. On 3 November, the regiment reported itself at 45 percent strength. Chinese losses probably were also as high as 600.[26]

The battle at Unsan exemplifies the CCF tactical style. The attack plan was based on thorough reconnaissance, accurate anticipation of the U.S. response, and a well-developed appreciation for the terrain. Stealthy movement permitted the Chinese to silently surround the Americans and achieve partial tactical surprise. Pressing the attack at night, the Chinese rapidly infiltrated the gaps and open flanks in the U.S. lines and established blocking positions on the escape routes to the rear. In addition, the bold Chinese crossing of the bridge over the Nammyon River demonstrated their confidence and cunning knowledge of their foe.

The Chinese published a pamphlet after their victory entitled "Primary Conclusions of Battle Experience at Unsan." The pamphlet cited the CCF's superiority over the U.S. Army in soldiering:

> Cut off from the rear, they [the Americans] abandon all their heavy weapons.... Their infantrymen are weak, afraid to die, and have no courage to attack or defend. They depend always on their planes, tanks, artillery.... They specialize in day fighting. They are not familiar with night fighting or hand-to-hand combat. If defeated, they have no orderly formation. Without the use of their mortars, they become completely demoralized. They are afraid when the rear is cut off. When transportation comes to a standstill the infantry loses the will to fight.[27]

The pamphlet also emphasized the use of the open V-formation to surround the enemy and the rapid infiltration of the enemy lines in order to slash through to the rear to block escape routes and prevent the advance of relief forces. In addition, it pointed out the value of using stealthy nighttime approaches to achieve surprise.

The battle at Unsan is only one of many operations that illustrate Chinese methods of offensive operations. The CCF attack at Chosin Reservoir against the U.S. Marine 1st Division is another good example of Chinese surprise, infiltration, envelopment, flank and rear attacks, operations at night, stealth, and establishment of road blocks. The great majority of CCF small-unit attacks also possessed these features.

The Defense

Chinese methods of defense basically took two forms corresponding to the two broad phases of the war: the mobile, maneuvering phase from the autumn of 1950 to the autumn of 1951 and the World War I-style tactical stalemate that existed after the autumn of 1951. The defense practiced during the first phase of the war was described by the G2 of IX Corps, Eighth Army, thusly:

> Interrogation of Prisoners of War and a study of captured enemy documents, as well as experience gained in action against the Chinese Communist Forces, have revealed that the underlying difference in concept between UN defense and Chinese Communist defense is that the one system depends upon strong defensive positions with supporting artillery and air cover, while the other, for lack of supporting arms, relies on a more fluid defense which actually takes the form of maneuvering tactics.[28]

During the initial period before battle lines became fixed in Korea, the CCF did not employ the principle of a main line of resistance or a position defense. Instead, they employed a basic defensive scheme of "one up and two back." In this scheme, the "up" group operated as a screening and delaying force. The two "back" units—out of artillery range—rested, regrouped, restocked, and reorganized for a counteroffensive or the defense. In the meantime, the screening elements conducted low-level limited attacks to confuse the enemy. If faced with a determined UN attack, the screening force offered stiff resistance but did not become decisively engaged. As it fell back slowly, contact was eventually made with the rearward units. If the rearward units still were unprepared, they, too, began a slow withdrawal until more favorable circumstances existed. During this stage of the war, the CCF usually preferred to continue to fall back under pressure until it could launch a counteroffensive rather than to stand fast on a predetermined line.[29]

A number of defensive principles characterized Chinese operations during this time:

● Defensive units, disposed in great depth, deployed along a narrow front.

● Forward elements played purely delaying roles to gain time while the remaining units prepared a second line of defense.

● Troops built defensive positions strong enough to afford protection from air and artillery attack.

● Soldiers established dummy positions and gun emplacements for the deception of the enemy.

● The Chinese placed light automatic weapons well forward, with the heavy weapons disposed in depth. Troops used heavy weapons primarily in support of a counterattack and fired mainly at night in order to avoid detection by UN air and ground observers.

● Defensive forces were withdrawn to successive defensive positions during hours of darkness only.[30]

Even under these temporary, mobile conditions, however, the Chinese constructed formidable defensive positions, as described by the IX Corps G2:

> (1) An investigation of one CCF position overrun by UN forces revealed 1,120 one-man foxholes, 684 two-man foxholes, 253 three-man foxholes, and 17 pillboxes, all of which could accommodate an estimated 3,250 men. These entrenchments were well camouflaged by logs covered with earth and were well protected against air attack by being positioned behind rocks and trees. The pillboxes were constructed of logs, dirt, and stone. These emplacements afforded maximum protection against mortar and small arms fire, but could be effectively neutralized by artillery or napalm.

(2) The examination of an enemy battalion defense position revealed that the emplacements were well dug-in and organized to a depth of approximately 2,000 meters. Fields of fire covered the slopes and draws and appeared to be well coordinated. A large quantity of ammunition of all types was found at the positions. Weapons and ammunition discovered included: 2 Japanese knee mortars, 6 Bren guns, 12 BARs, 5 U.S. light machine guns, 2 U.S. heavy machine guns, 30—40 U.S. M-1 rifles, 30—40 U.S. carbines, many rifles of foreign make, a large amount of assorted ammunition including 2,000 hand grenades of the potato masher type. The command post was well dug-in on the reverse slope of the hill. Bunkers were well constructed with over-head cover.[31]

This report is also noteworthy in that it reveals how much the Chinese relied on captured U.S. weapons and ammunition and how diverse the assortment of materiel was.[32]

During the last two years of the war, the Chinese defense assumed a positional character of remarkable strength. By the end of 1951, the extensive trench network ran fourteen miles in depth.[33] As time passed, the works became more and more impregnable. By hand labor, using ordinary tools, CCF troops fortified the reverse slopes of hills and dug tunnels all the way through to the forward slopes for observation.[34] Furthermore, entire units were housed underground with only observers left above ground.[35]

Placing the main line of resistance underground on the reverse slope reduced the vulnerability of the CCF to observation and direct and indirect fires. Moreover, it permitted the Chinese to support each rear hillside or back ridgeline (and some forward slopes) with supporting fires from adjacent high ground. In the attack, UN forces had to deploy to clear observers and small combat elements from the forward slope, all the while taking fire from enemy mortars and artillery pieces. Once on the crest and descending, the UN forces lost the beneficial effects of their own artillery support and fell victims to heavy direct fire from hidden enemy positions, many of which did not become evident until killing rounds burst out of them. If forced to retreat, the UN forces then had to fall back through the enemy indirect fires one more time. Clearly, reverse-slope defenses have deadly effectiveness for light forces when they are properly constructed and coordinated.

CCF defensive works exploited the terrain fully and followed an irregular shape, often triangular or ladderlike, so that rearward positions could fire in the gaps between the forward positions. Fighting positions lay behind trees, hedges, and natural rock outcrops. Earth mounds conformed to and molded with the existing contours of the land. Fortifications retained a low silhouette so as not to stand out on the skyline. Communication trenches, often covered, connected the most important weapons positions and led back to switch positions.[36]

The Chinese constructed their fortifications in such a way as to maximize flanking fire, especially by their machine guns, which they considered to be the backbone of infantry firepower. Obstacles covered by fire and observation were placed to the front and flanks to channel the enemy into the fire lanes. The Chinese used mines extensively; most of them were captured from UN stocks or improvised with explosives using a wide variety of containers: glass jars, clay pots, tin cans, wooden boxes, and fuel drums.[37]

Camouflage was essential, and the Chinese observed meticulous camouflage discipline. The CCF also added to deception through the use of dummy positions. A Chinese reference manual on field fortifications made several good points on how to incorporate dummy positions into the defensive lines:

> When a structure cannot easily be concealed perfectly, construct many dummy structures so that the enemy will not be able to distinguish the real from the false. These dummy structures will draw enemy fire, disperse enemy fire, and cause them to misuse their forces.
>
> The dummy structure should not be too close to the true structure lest it draw enemy fire to the true structure. Moreover, it should not be allowed to fall into enemy hands.
>
> The dummy structure should also be camouflaged and should sometimes be equipped with dummy soldiers and weapons. . . . The dummy structure need not be perfectly identical. . . . It is only necessary that it agree with the true structure in outward appearance.[36]

This Chinese manual is quite detailed. It describes how to construct and integrate positions for personnel and weapons of every kind—artillery, antitank guns, even horses. It specifies the man-hours and tools required for the construction and provides several hundred diagrams. A few representative examples of these diagrams are at appendix A to this chapter. Building and camouflaging these fortifications required enormous labor.

Despite attacks by unprecedented levels of fire by aircraft and artillery of all calibers (which was the primary U.S.-UN response to Chinese defenses),

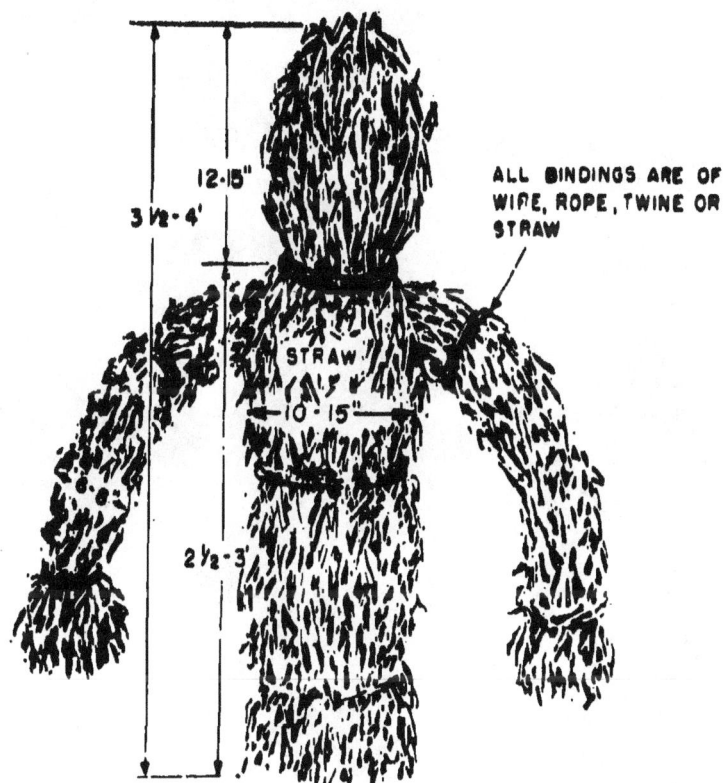

"Strawman" decoy

Chinese positions usually had to be cleared by close-in assault, frequently using flamethrowers. The main defensive positions, however, continued to be screened by strong outposts that could be pushed in by heavy patrolling.[39]

Tactically, the Chinese fought as tenaciously in the defense as they had in the attack. To conserve ammunition and remain undetected, they held their fire until they were certain of its effectiveness, then opened up with a withering volume of small-arms fire and grenades. Longer-ranged crew-served weapons were employed according to strict distance limitations: 60-mm mortars from 1,000 to 1,500 meters and machine guns from 300 to 500 meters.[40]

Chinese Communist POWs, wearing quilted-cotton winter uniforms and fleece-lined caps

73

To augment their defenses, the Chinese increased the heaviness of their forces. The Chinese added tank, antiaircraft, and armored car units, reequipped and increased their infantry, and built up their air strength.[41] More artillery and heavy-mortar units were also obtained. Artillery and mortar positions were dug in up to six feet in depth and positioned, often by hand, in unlikely places. Ammunition was similarly revetted and camouflaged. Although the use of indirect fires increased, artillery techniques and effectiveness still remained below that of the UN Command.

Counterattacks also filled an important place in Chinese defensive doctrine. Counterattack tactics generally conformed to the Chinese offensive doctrine described earlier. Following thorough reconnaissance and infiltration, small-unit attacks were conducted at night against the UN flanks and rear. Counterattacks were sometimes conducted as spoiling attacks or to blunt the edge of a UN offensive. The Chinese also conducted immediate counterattacks to retake lost positions. These counterattacks had to be launched early enough in the evening to leave several hours of darkness for the Chinese to repair defensive works by morning.

In their defenses, the Chinese made heavy use of booby traps. Field-expedient and improvised mines and booby traps (for example, dud bombs and mortar shells) proved to be the rule.[42] The CCF used these sorts of defenses imaginatively. In one case, Chinese troops buried a mortar shell under the ashes of a burned-out fire and placed a small amount of fresh wood on the heap. This inviting sight prompted UN personnel in the area to build a warming fire. The mortar bomb exploded two hours after the fire was lighted.[43] The Chinese frequently booby-trapped logs and branches (which were scarce) expecting the enemy to use them in building fires. In another case, the CCF draped a large wire entanglement across a village thoroughfare. Various ends of the wire were connected to hidden grenades. Any attempt to pull the wire out resulted in an explosion. The Chinese were also known to booby-trap dead bodies. UN personnel had to stay alert.

Shattering this tough defensive barrier proved to be a very difficult task. Long, bloody battles, fought over small pieces of ground, found their way into historical lore and legend: Heartbreak Ridge, Pork Chop Hill, Bloody Ridge. In the end, the UN Command—and the United States, in particular—was unwilling to apply the manpower and suffer the casualties necessary to punch through the Chinese defenses and drive them from Korea. Thus, the war ended with two armed camps firmly entrenched and facing each other across a no-man's land of battle-scarred terrain.

Logistics

Had the CCF possessed a modern, well-organized, efficient logistic system comprising motor transport and stocks on the same scale as the United States, the Eighth Army probably would have been annihilated in Korea in the fall and winter of 1950. Instead, the primitive, unreliable logistics of the CCF did not permit it to continue an offensive beyond a period of three or four days. Thus, the CCF could not exploit tactical successes in depth due to its inherent lack of mobility and sustainability. Assaults that required days of buildup, even when they were accompanied by spectacular successes, lost impetus in just a few days. The Chinese leadership itself acknowledged that poor supply was their greatest difficulty during the war.[44]

Three fundamental weaknesses crippled the Chinese logistic system. The first weakness was organizational in nature. Operating from a poor economic base, they simply did not have the stores of military supplies (particularly ammunition) nor the transport necessary to sustain their large army over long distances.[45] Second, the portion of Korea under Chinese control was barely able to sustain itself, much less meet the needs of hundreds of thousands of foreign troops. Finally, the entire length of the CCF lines of communication was under constant attack by UN airpower. Depots, truck columns, railroads, trains, transportation junctions, tunnels, and bridges were destroyed time and again by UN aircraft. As a result, Chinese supplies moved almost exclusively at night. Under these debilitating conditions, the Chinese survived by virtue of improvisation, discipline, and sheer perseverance.

Firewood booby trap

The primary task of the Chinese rear services, as part of their logistic system, was to keep supplies moving forward. Supplies originating in Chinese Manchuria moved by night in trains and truck columns to forward army depots—the trains sheltering in tunnels for protection during the day. Supply points were well camouflaged and protected. Where possible, local supplies were made available to Chinese units. In support of the CCF's rear services, the North Korean government organized more or less permanent repair teams to rebuild bridges, tunnels, railroad lines, and roads immediately following their damage by air strikes. Prodigious support by North Korean natives also helped immensely. In one instance, thousands of local peasants restored 180 miles of road to truck use within 36 hours. In another case, a 37-foot-high, 150-foot-long ramp composed of earth in rice bags was built to link one end of a blown-up bridge with the near bank.[46] Naturally, assistance like this enabled the CCF to keep its soldiers in forward foxholes, not in rear areas.

At division level and below, the CCF used the resources at hand to feed and move itself. Supplies moved solely by human and animal muscle power. Groups of Korean porters, under Chinese guard, were organized to carry unit provisions forward and even into combat. Oxcarts, camels, and ponies hauled materiel over the restrictive terrain by night. Each soldier began an offensive with a heavy load: 3-days' rations, his bedroll, 4 grenades, 100 rounds of ammunition, and a mortar bomb or 2. The Chinese procured some of their food locally, sometimes by force, sometimes by legitimate means. At times, they required villagers to cook for them. Captured UN supplies were also a ready source of ammunition, equipment, and rations; in many cases, the Chinese replenished their stocks after a successful attack. The Chinese also buried supplies when withdrawing from an area with the expectation that the caches would be dug up and used upon their return.

In the worst conditions, the CCF soldier learned to do without. His self-discipline led him to subsist on meager rations and to forego nonessentials.

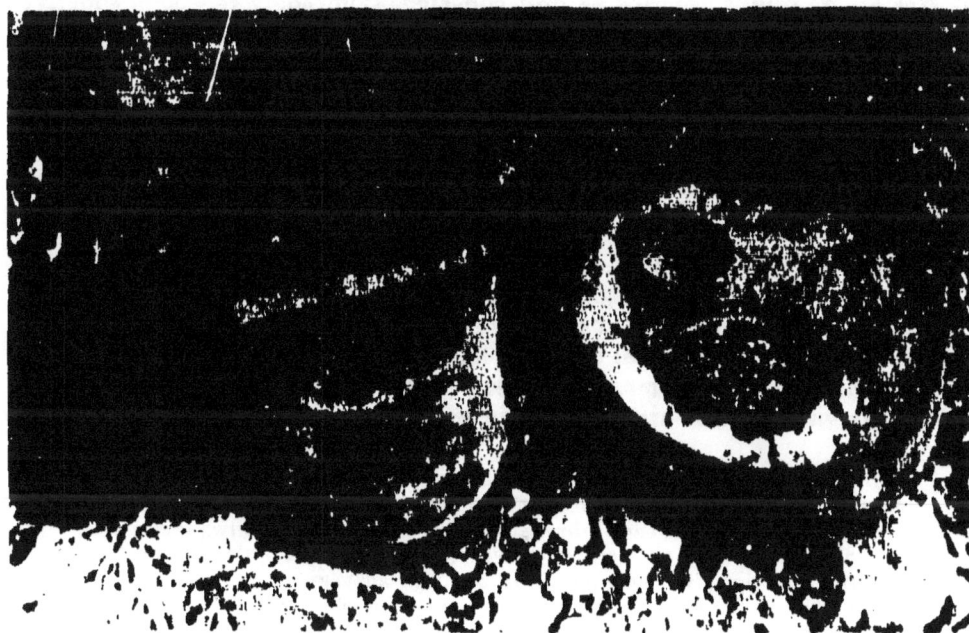

Meager Chinese rations

76

In combat, the Chinese infantryman also learned to pass up promising targets in order to conserve his ammunition for critical times. In fact, there are a few accounts of Chinese soldiers going into the attack without firing their weapons at all. The stoicism, perseverance, and hardiness of CCF light infantrymen stood them well during hard times.

CCF Leadership

How was the CCF able to accomplish as much as it did during the Korean War given the woeful inadequacies of its logistics and the overwhelming superiority of UN firepower? As mentioned earlier, part of the answer lay in the CCF's philosophy of "man over weapons." The application of this philosophy obtained maximum value from the CCF by focusing on its most potent capability—human will. By sheer efforts of will, CCF infantrymen were able to rise above their weaknesses in materiel. The Chinese leadership was able to successfully mobilize the superior human element in its men. Thus, leadership was crucial to the effective small-unit actions that were so critical to CCF operations.

To ensure effective leadership, Chinese combat leaders operated through a number of institutional structures and techniques. So that control would be assured, military leaders had to be true believers. Thus, virtually all cadres were dedicated Communists in good standing. In their person, they represented the establishment and had a personal stake in the official policy, doctrine, and objectives. They were dedicated to the Chinese involvement in the war and to specific CCF methods.[47]

One of the objectives of leaders was to establish "comradely relations" as the basis for actions on the battlefield.[48] Comradely relations went far beyond what Westerners might describe as unit cohesion. Comradely relations sought the total dedication of individual soldiers through their involvement in the small-group life of a unit—a group life which approached the intensity of a military-religious order. This philosophy incorporated the principles of solidarity, political loyalty, fierce determination, and the ethical responsibility to fight on and endure. Individualism was ruthlessly suppressed in favor of group identity. Soldiers were made to believe that their well-being and survival depended entirely on the small group.

To instill these principles, the CCF leadership used such means as political conversion, indoctrination, and egalitarianism (in terms of uniforms, privileges, and polite forms of address among all ranks). Perhaps the most important technique was an organizational one—the 3x3 cellular organization established within squads.[49] General James Van Fleet, commander of the Eighth Army, described the value of this arrangement in this way:

> The Red Chinese Army is divided at the very bottom into units of three men, with each assigned to watch the others and aware that they in turn are watching him. Even when one of them goes to the latrine, the other two follow. No soldier dares fail to obey orders or even complain.... The little teams of three, each man warily watching the others, begin the advance.... Yet—although terribly alone in the fight despite the two men at his side, made even more lonely by the doubt whether the two are there to help him or to spy on him— the Red soldier moves ever forward....[50]

77

Certainly, this account is oversimplified, yet it points out the importance of this controlling device in creating conformity and motivation in the CCF.

This sort of tight organization also facilitated command and control and gave the squad leader three tidy combat groups to use in rapid tactical responses. Its effectiveness for light infantry operations has been of high interest to some Western officers.

Although CCF leaders, particularly the political commissars who were assigned down to company level, attempted to manipulate and control their men, they also showed a true regard for their welfare. Care was taken not to institute arbitrary or harsh discipline. Soldiers apparently had the right to raise legitimate complaints without fear. Furthermore, through precombat briefings, the men were led to feel as though they were participants in the decision-making process; they were more likely then to fight out of loyalty than duty or fear. Group meetings were held during which soldiers were exhorted and encouraged to declare their loyalty to the group and to take oaths. An awards system cultivated soldierly honor and raised soldier prestige. Moreover, the leaders explained to the men why they were in Korea and what they hoped to accomplish, stressing their superior moral position vis-a-vis the UN command.[51]

The Chinese cadre also led by personal example. In combat they were in the forefront, exhorting, motivating, and directing their men. In retreat, they were the last to fall back. Furthermore, they suffered the same privations as their men and exhibited courage and determination in all circumstances.

Through the means of unrelenting group pressure, strict organizational controls, moral and political indoctrination, individual co-optation, and personal example, CCF leaders forged the "comradely relations" necessary to execute the particular tactical style of the CCF. The effectiveness of these methods of leadership, command, and control is borne out by the outstanding tactical performances of the CCF small units.

Summary and Conclusions

The main strengths of the CCF in the Korean War were the power of the philosophy of "man over weapons," the skills and abilities of the individual light infantrymen, and the effectiveness of the CCF leadership. The integration of these strengths created a fierce battlefield instrument that achieved remarkable tactical successes, even while hampered by crippling weaknesses.

The Chinese leadership's emphasis on the superiority of its soldiers and its assertion that UN advantages in materiel and weapons were insignificant created confidence in the infantry ranks that they could defeat the UN command. Their confidence proved well founded, at least in the first year of the war, when the Chinese frequently demonstrated superior field craft, almost inhuman endurance, and a sharp appreciation for terrain. Undaunted by weather, terrain, or privation, the CCF, during this stage of the war, pressed the UN Command to its limits. Eventually, however, as the nature of the war changed—particularly as UN lines firmed up and were tied in—these Chinese strengths were nullified.

Another Chinese strength was their marked ability to improvise. The CCF used whatever resources were at hand for the military purposes of camouflage, deception, booby traps, fortifications, and sustenance. The record of CCF operations in the Korean War is one of resourcefulness, of using ingenuity to compensate for lack of materiel. Many U.S. tactical after-action reports note this uncanny Chinese propensity for improvisation.

The most obvious weakness of the CCF was its severe shortage of military equipment for combat support and combat service support. Furthermore, the Chinese were hopelessly outmatched by the UN in firepower, transport, and airpower. The CCF never had enough artillery, trucks, aircraft, signal equipment, medical equipment, or combat stores to support its infantry armies. The Chinese logistic system was also a major weakness. Its inability to sustain offensives beyond three or four days is well documented. Of course, the crushing effect of U.S.-UN airpower and long-range artillery on Chinese lines of communication must not be overlooked.

A further Chinese debility was their tactical rigidity. This weakness characterized CCF patrolling, in that flank and rear guards were not used, and routes were reused even though patrols showed repeated vulnerability to ambushes. Tactical rigidity was also the result of the Chinese lack of signal equipment. Lacking adequate communications, the Chinese maintained attacks even when outcomes appeared hopeless, thus taking excessive casualties.

These weaknesses were all magnified during the last two years of the war. Once the UN Command had established a solid defensive line from coast to coast backed by huge volumes of indirect fires and airpower, Chinese short-comings proved more damaging. Furthermore, the Chinese advantages in tactical maneuver, infiltration, and stealth lost their value. The CCF was no longer able to take objectives by slipping through thin lines to attack the enemy flank or rear. The Chinese occasionally conducted human wave attacks out of frustration with this situation.[52]

Ultimately, the CCF suffered an erosion of morale. By maintaining unchallenged command of the sea and the air, inflicting continuous damage to lines of communication, and delivering shocking bombardments against Chinese line units, UN forces, through their technical superiority, finally asserted their massive advantages.

By the autumn of 1951, the CCF leadership could no longer deny that its deficiencies in materiel doomed it to a tactical stalemate at best. Realizing its impotence, the CCF lost its psychological advantage over the UN forces and began to suffer a morale problem. UN firepower had equalized the manpower imbalance and, in the final analysis, negated Chinese strengths. Thus, the Korean War represents the limits to which the "man over weapons" philosophy can be carried.

Finally, it is important to reiterate that, aside from their initial strategic intervention, Chinese light infantry armies *could not* operate at the operational level of war. Deficiencies in long-range weapon systems, sustainability, and transport prohibited the development of a capability for deep maneuver by the CCF. When coupled with the devastating deep interdiction of UN air forces and the lack of maneuver space, these deficiencies imposed a tactical ceiling on CCF operations. Even though the CCF offensives of 1951 involved several armies of hundreds of thousands of men, they assumed a tactical character, albeit on a huge scale. The fitful start-and-stop pattern of attack, regroup, restock, and attack limited the CCF to a series of short-range tactical successes that were eventually blunted by the firepower and defenses of the UN Command.

Appendix A*
Illustrations of CCF Fortifications

Illustration 50. Emplacement for 57 mm anti-tank gun.

Cross section view from A-B   Top view

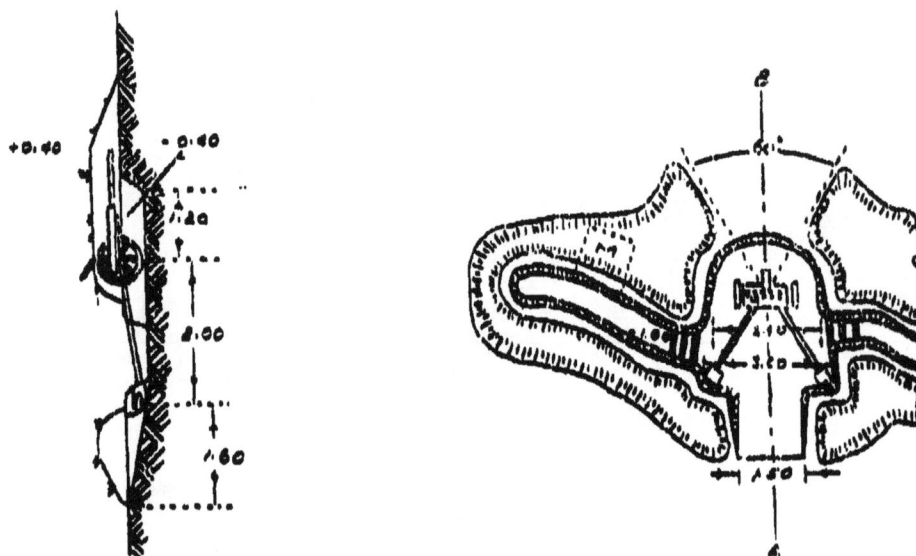

Note: Amount of dirt excavated: 5.65 cu m.
Time of completion: 9.49 man hours.

*Source: U.S. Army, Corps of Engineers, Intelligence Division, "Enemy Field Fortifications in Korea," no. 15, in *Engineer Intelligence Notes* (Washington, DC: Army Map Service, January 1952), 2–8; *Chinese Communist Reference Manual for Field Fortifications*, translated by the Military Intelligence Section, General Staff, Far East Command, 1 May 1951, 63, 64, 112, 178.

Illustration 51.
Emplacement for artillery.

Figure 1: A simple one

Trench for men   Trench for ammunition

Figure 2. A complete one: A-B a primary line

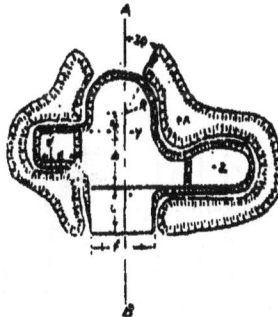

Note 1. Amount of dirt excavated and time required for completion:

Type of weapon	Amount of dirt excavated	Time of completion
75 mm Howitzer	6.85 cu m	17.81 man hours
Field gun	20.62 cu m	53.61 man hours
Mountain gun	13.08 cu m	36.09 man hours
150 mm Howitzer	45.44 cu m	118.14 man hours
100 mm Cannon	43.20 cu m	112.32 man hours

333. Excavated type of light shelter.

The excavated type of light shelter is dug simultaneously with the trenches, and then the cover is placed over it. In case of necessity, the cover should be erected first, then the shelter dug after the trenches are completed.

Illustration 101. Excavated
type of light shelter.

1. Plans view.

2. A-B Cross section.

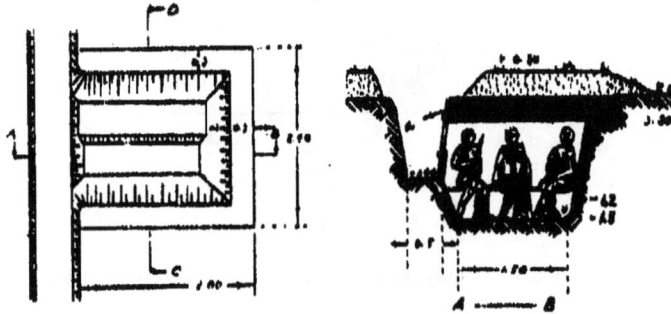

a. The door board is used to protect personnel against enemy shrapnel and bullets.

C-D Cross Section

a. Wooden tie.

b. Cover timber.

Illustration 104. Reverse
Slope Light Shelter (gentle slope).

1. Flank view.

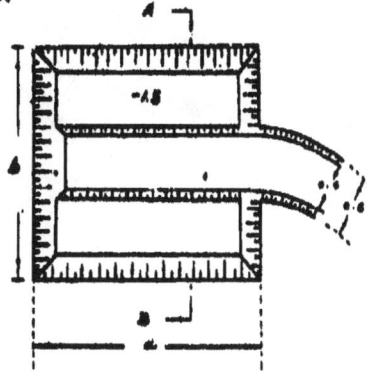

a. Approximately 2.00
b. Varies according to topography.

2. A-B Cross section.

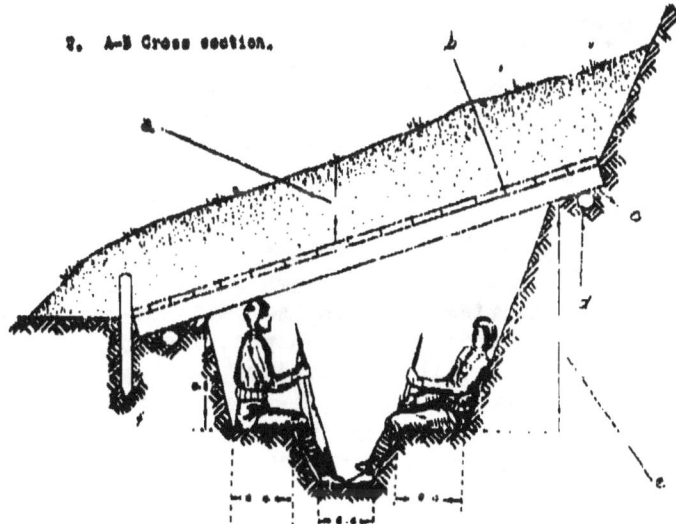

a. Approximately over 50 cm
b. Cover board
c. Rafter
d. Wooden tie
e. Approximately 1.8
f. Base board

118

84

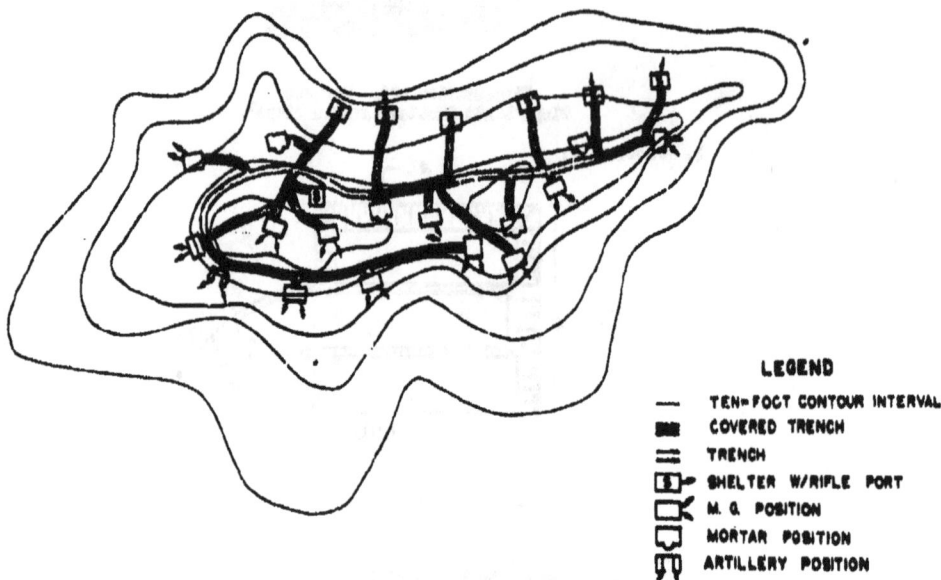

LEGEND

— TEN-FOOT CONTOUR INTERVAL
▬ COVERED TRENCH
= TRENCH
⊡ SHELTER W/RIFLE PORT
⊏ M. G. POSITION
⊓ MORTAR POSITION
⋔ ARTILLERY POSITION

Fig. 1. Typical hill defense system.

2. Types of Emplacements

a. Trenches -- Trench systems (Figure 2) are extensive and well-laid-out on the enemy-defended hills of Korea. Each hill has one main communication trench following the contour of the reverse slope. From the main trench, short connecting trenches branch off to emplacements and shelters.

The main trench has heavy overhead cover at short intervals; it also has small-arms positions and 1-man shelters cut into its walls. In most cases, the connecting trenches are well-covered; they are tunneled wherever possible, especially between positions on the reverse and forward slopes (Figure 3). All the trenches, average 5 to 6 feet in depth and 1½ to 2 feet in width. The overhead cover for the trenches is formed by a 3- to 6-foot layer of logs and earth. The tunnels are not dug to any standard depth below the surface. They are generally

2

85

2 feet wide by 3 feet high, although some are only 2 feet square. All the tunnels are shored with timber, wherever necessary.

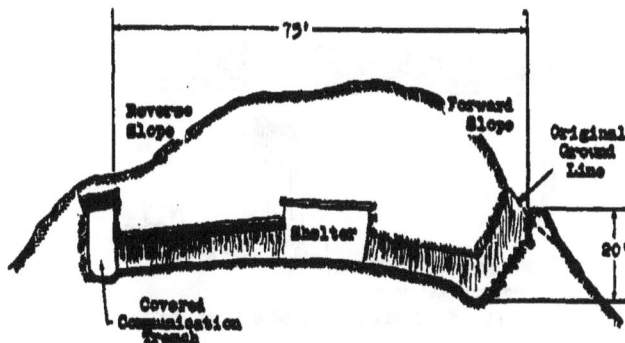

Fig. 3. Tunnel between forward and reverse slope positions.

b. Rifle Positions -- Individual rifle positions are located on both the forward and reverse slopes for all-round defense (Figure 4). In some cases, three or four positions may be interconnected by tunnels, especially where a sharp ridge line exists to make extensive tunneling unnecessary.

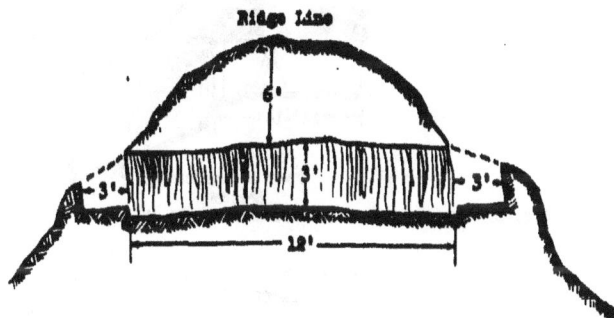

Fig. 4. Individual rifle positions connected by a tunnel.

c. Troop Shelters -- Troop shelters have no standard size. They are normally built on reverse slopes and in many cases they serve as alternate firing positions. These shelters have a capacity of two to eight men, and have a headroom of only 4 to 5 feet.

The overhead protection of these shelters ranges in thickness from 3 to 12 feet and consists of many layers of logs and a cover-layer of earth. Logs 4 to 10 inches in diameter have been found placed in the overhead protective cover. Logs up to 13 inches in diameter serve as support posts. A cross section of a typical troop shelter is shown in Figure 5.

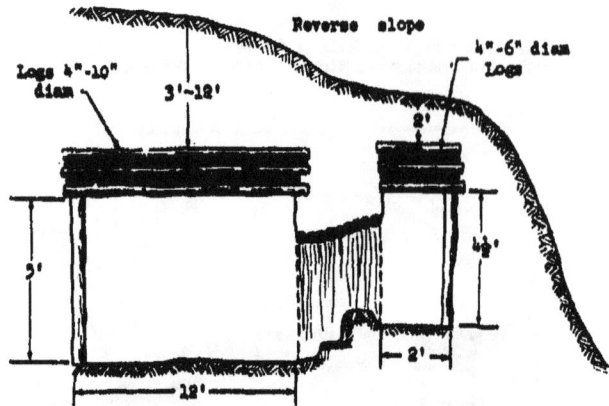

Fig. 5. Cross section of a troop shelter.

d. Mortar Emplacements -- Where the terrain permits, mortar emplacements are usually sited on the reverse slopes. Occasionally, they may be found on the forward slopes. The emplacements (Figures 6 and 7) are dug about 4 feet deep and provided with overhead

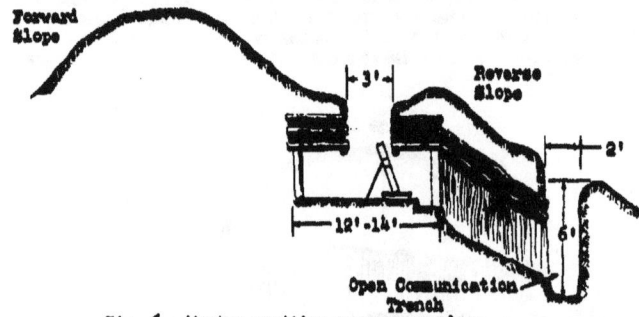

Fig. 6. Mortar position on reverse slope.

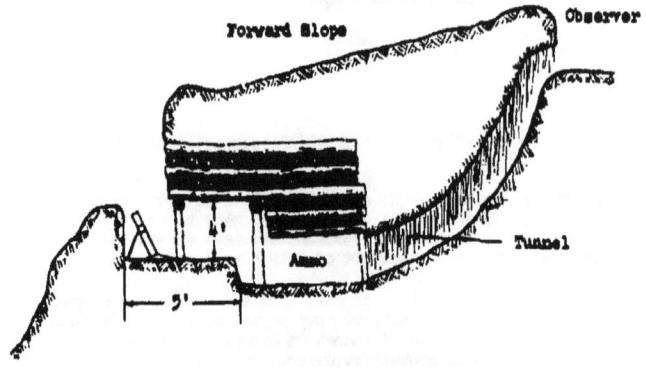

Fig. 7. Mortar position on forward slope.

cover for the crew. Most mortar positions are sited to cover dead areas in the field of fire of flat trajectory weapons on the forward slopes.

As an example of diverse materials used in construction, one mortar emplacement was found with an overhead cover formed by a piece of sheet iron. The mortar was fired through a square opening in the sheet iron, which, however, offered less protection than the conventional log-and-earth covering.

e. Machine Gun and Automatic Weapon Emplacements -- These types of emplacements are quite numerous; wherever possible they are positioned in depth along the forward slopes of hills and their crests (Figure 8). They are the ordinary cut-and-cover type of emplacements, with the emphasis on cover.

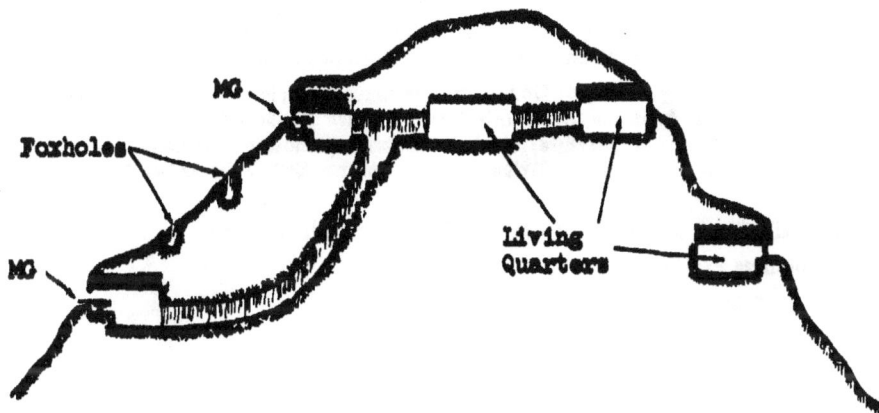

Fig. 8.   Cross section of hill, showing machine-gun emplacements and shelters.

NOTES

Chapter 2

1. S. L. A. Marshall, *The Military History of the Korean War* (New York: Franklin Watts, 1963), 5.

2. Matthew B. Ridgway, *The Korean War* (Garden City, NY: Doubleday and Co., 1967), 4.

3. Alexander L. George, *The Chinese Communist Army in Action: The Korean War and Its Aftermath* (New York: Columbia University Press, 1967), vii.

4. Figures 3–5 and table 3 are taken from an unidentified publication by the U.S. Army Far East Command, reproduced by the 8218th Engineer Topographic Detachment. These materials were filed loose in U.S. Army, Forces in the Far East, "Chinese Communist Ground Forces in Korea, Tables of Organization and Equipment," 1953.

5. Marshall, *Korean War*, 33.

6. U.S. Army, IX Corps, G2, *Enemy Tactics, Techniques and Doctrine* (Seoul, Korea?, 24 September 1951), 4. Hereafter cited as *Enemy Tactics, Techniques and Doctrine*.

7. Ibid., 6. See also U.S. Army, Corps of Engineers, Intelligence Division, "Enemy Camouflage Practices in Korea," no. 8, in *Engineer Intelligence Notes* (Washington, DC: Army Map Service, September 1951), 1–4, hereafter cited as "Enemy Camouflage Practices in Korea." The Corps of Engineers, Intelligence Division author information is hereafter cited as CE, IQ.

8. S. L. A. Marshall, *The River and the Gauntlet* (New York: William Morrow and Co., 1953), 9.

9. "Enemy Camouflage Practices in Korea," 4.

10. *Enemy Tactics, Techniques and Doctrine*, 5.

11. Ibid., 52.

12. Ibid., 10.

13. Ibid., 9.

14. S. L. A. Marshall, *Commentary on Infantry Operations and Weapons Usage in Korea, Winter of 1950–51*, Project Doughboy, Report no. ORO-R-13 (Chevy Chase, MD: Operations Research Office, Johns Hopkins University, 1952), 5.

15. Marshall, *River and the Gauntlet*, 58–59.

16. CE, IQ, "Stream-Crossing Expedients of NKPA and CCF Foot Troops," no. 10, in *Engineer Intelligence Notes* (Washington, DC: Army Map Service, October 1951), 1–4. See also *Enemy Tactics, Techniques and Doctrine*, 7. For an actual tactical account of Chinese techniques in crossing shallow rivers, see chapter 3, "The Affair at Chinaman's Hat," in Marshall, *River and the Gauntlet*, 41–55.

17. U.S. Army, 8th Army (Korea), Historical Section, *Special Problems in the Korean Conflict and Their Solutions* (Seoul, Korea?, 1952), 105–6.

18. George, *Chinese Communist Army*, 3.

19. Lynn Montross and Nicholas A. Canzona, *U.S. Marine Operations in Korea, 1950–1953*, vol. 3, *The Chosin Reservoir Campaign* (Washington, DC: Historical Branch, G-3, U.S. Marine Corps, 1957; St. Clair Shores, MI: Scholarly Press, 1976), 92.

20. *Enemy Tactics, Techniques and Doctrine*, 17–20. Most of the text and figure 8 in this section are drawn from this document.

21. Marine Colonel A. L. Bowser as quoted in Montross and Canzona, *Marine Operations . . . Chosin Reservoir*, 92.

22. U.S. Army, 8th Army (Korea), *Enemy Tactics* (Seoul, Korea?, 1 November 1952), 12A, 48A.

23. *Enemy Tactics, Techniques and Doctrine*, 37—38. A CCF division was normally supported by only one artillery battalion. Close support during an attack generally did not occur. Infantry requests for fire first went through the infantry battalion commander before entering artillery channels.

24. Ridgway, *Korean War*, 186.

25. Marshall, *Commentary*, 99.

26. The preceding account of the Battle of Unsan is summarized from Roy E Appleman, *South to the Naktong, North to the Yalu*, United States Army in the Korean War (1961; reprint, Washington, DC: Office of the Chief of Military History, Department of the Army, 1973), 689—708.

27. T. R. Fehrenbach, *This Kind of War: A Study in Unpreparedness* (New York: Macmillan Co., 1963), 300—301.

28. *Enemy Tactics, Techniques and Doctrine*, 22.

29. Ibid.

30. Ibid., 23.

31. Ibid.

32. The CCF entered Korea already armed with a large collection of U.S. and Japanese weapons captured during World War II and the Chinese Civil War.

33. Marshall, *Korean War*, 70.

34. Ridgway, *Korean War*, 186.

35. Marshall, *Korean War*, 70.

36. *Chinese Communist Reference Manual for Field Fortifications*, translated by the Military Intelligence Section, General Staff, Far East Command, 1 May 1951, 5—11. Also, CE, IQ, "Enemy Field Fortifications in Korea," no. 15, in *Engineer Intelligence Notes* (Washington, DC: Army Map Service, January 1952), 1—7.

37. CE, IQ, "Chinese Communist Mine Warfare," no. 5, in *Engineer Intelligence Notes* (Washington, DC: Army Map Service, April 1951), 1—7; and CE, IQ, "Enemy Improvised Mines in Korea," no. 18, in *Engineer Intelligence Notes* (Washington, DC: Army Map Service, August 1952), 1—11.

38. *Chinese Communist Reference Manual*, 3—4.

39. U.S. Army Field Forces, "Dissemination of Combat Information—Essential Reports," (1951), 3—4.

40. *Enemy Tactics, Techniques and Doctrine*, 23.

41. George, *Chinese Communist Army*, 199.

42. *Enemy Tactics, Techniques and Doctrine*, 15; and CE, IQ, "Booby Traps Employed by the NKPA and CCF," no. 12, in *Engineer Intelligence Notes* (Washington, DC: Army Map Service, December 1951), 1—5.

43. Ibid.

44. U.S. Army, I Corps, G2, "CCF Logistical Capabilities: A Study of the Enemy Vehicular Effort on I Corps Front" (Seoul, Korea?, 28 June 1952), 15. See also U.S. Military Academy, West Point, Department of Military Art and Engineering, *Operations in Korea* (West Point, NY, 1953), 14, 37.

45. Ibid., 2.

46. CE, IQ, "Chinese Communist Engineers," no. 19, in *Engineer Intelligence Notes* (Washington, DC: Army Map Service, November 1952), 10; CE, IQ, "Military Construction Practices by the Enemy in Korea," no. 16, in *Engineer Intelligence Notes* (Washington, DC: Army Map Service, March 1952), 3; and CE, IQ, "Railroad Repair and Reconstruction by NKPA and

CCF," no. 17, in *Engineer Intelligence Notes* (Washington, DC: Army Map Service, June 1952), 1—11.

47. George, *Chinese Communist Army*, 53.

48. Ibid., 27.

49. Ibid., 51.

50. Ibid., 52.

51. Ibid., 127—59. The author stresses the strong, favorable impact of political indoctrination and precombat briefings on the morale and sacrificial attitudes of CCF troops.

52. Gerard H. Corr, *The Chinese Red Army: Campaigns and Politics Since 1949* (New York: Schocken Books, 1974), 89.

- - - - - - - - - - - - - - - - - - -

BIBLIOGRAPHY

Chapter 2

Appleman, Roy E. *South to the Naktong, North to the Yalu.* United States Army in the Korean War. 1961. Reprint. Washington, DC: Office of the Chief of Military History. Department of the Army, 1973.

Chinese Communist Reference Manual for Field Fortifications. Translated by Military Intelligence Section, General Staff, Far East Command, 1 May 1951.

Collins, Joseph Lawton. *War in Peacetime: The History and Lessons of Korea.* Boston: Houghton Mifflin Co., 1969.

Corr, Gerard H. *The Chinese Red Army: Campaigns and Politics Since 1949.* New York: Schocken Books, 1974.

Fehrenbach, T. R. *This Kind of War: A Study in Unpreparedness.* New York: Macmillan Co., 1963.

George, Alexander L. *The Chinese Communist Army in Action: The Korean War and Its Aftermath.* New York: Columbia University Press, 1967.

Gugeler, Russell A. *Combat Actions in Korea.* Rev. ed. Washington, DC: Office of the Chief of Military History, U.S. Army, 1970.

Hermes, Walter G. *Truce Tent and Fighting Front.* United States Army in the Korean War. 1966. Reprint. Washington, DC: Office of the Chief of Military History, U.S. Army, 1973.

Marshall, S. L. A. *Commentary on Infantry Operations and Weapons Usage in Korea, Winter of 1950—51.* Project Doughboy. Report no. ORO-R-13. Chevy Chase, MD: Operations Research Office, Johns Hopkins University, 1952.

————. *The Military History of the Korean War.* New York: Franklin Watts, 1963.

————. *The River and the Gauntlet.* New York: William Morrow & Co., 1953.

Montross, Lynn, and Nicholas A. Canzona. *U.S. Marine Operations in Korea, 1950—1953.* Vol. 3. *The Chosin Reservoir Campaign.* Washington, DC: Historical Branch, G-3, Headquarters, U.S. Marine Corps 1957; St. Clair Shores, MI: Scholarly Press, 1976.

Ridgway, Matthew B. *The Korean War.* Garden City, NY: Doubleday and Co., 1967.

U.S. Army. 8th Army (Korea). *Enemy Tactics.* Seoul, Korea?, 1 November 1952.

U.S. Army. 8th Army (Korea). Historical Section. *Special Problems in the Korean Conflict and Their Solutions.* Seoul, Korea?, 1952.

U.S. Army. I Corps. G2. "CCF Logistical Capabilities: A Study of the Enemy Vehicular Effort on I Corps Front." Seoul, Korea?, 28 June 1952.

U.S. Army. IX Corps. G2. *Enemy Tactics, Techniques and Doctrine.* Seoul, Korea?, 24 September 1951.

U.S. Army. Corps of Engineers. Intelligence Division. *Engineer Intelligence Notes.* Washington, DC: Army Map Service, 1951—1954.

Individual issues cited below:

No. 5. "Chinese Communist Mine Warfare." April 1951.

No. 8. "Enemy Camouflage Practices in Korea." September 1951.

No. 10. "Stream-Crossing Expedients of NKPA and CCF Foot Troops." October 1951.

No. 12. "Booby Traps Employed by the NKPA and CCF." December 1951.

No. 15. "Enemy Field Fortifications in Korea." January 1952.

No. 16. "Military Construction Practices by the Enemy in Korea." March 1952.

No. 17. "Railroad Repair and Reconstruction by NKPA and CCF." June 1952.

No. 18. "Enemy Improvised Mines in Korea." August 1952.

No. 19. "Chinese Communist Engineers." November 1952.

No. 23. "Demolition Equipment Employed by the CCA and NKA." March 1954.

U.S. Army. 1st Cavalry Division. *The First Cavalry Division in Korea, 18 July 1950—18 January 1952.* Atlanta, GA: Albert Love Enterprises, 195?.

U.S. Army. 3d Division. *3d Infantry Division in Korea.* Edited by Max W. Dolcater. Tokyo?, 1953.

U.S. Army Field Forces. "Dissemination of Combat Information—Essential Reports." 1951.

————. "Report of Army Field Forces Observer Team Number 6." 7 April 1952.

U.S. Army. Forces in the Far East. "Chinese Communist Ground Forces in Korea. Tables of Organization and Equipment." 1953.

U.S. Military Academy, West Point. Department of Military Art and Engineering. *Operations in Korea.* West Point, NY, 1953.

Whiting, Allen S. *China Crosses the Yalu: The Decision to Enter the Korean War.* New York: Macmillan Co., 1960; Stanford, CA: Stanford University Press, 1968.

Chapter 3

British Operations in Malaya and Borneo, 1948–1966

Part I. The Malayan Emergency

Introduction

From 1948 to 1966, substantial British, Gurkha, and Commonwealth infantry forces participated almost continuously in protracted light infantry operations in the Far East. In Malaya, from 1948 to 1960, these British-directed forces defeated an indigenous Communist insurgent force. Less than three years later, the British Army moved into North Borneo to secure that territory against Communist guerrillas and Indonesian aggression in a four-year war. In both wars, the combat took place in extremely inhospitable terrain and was swift, fleeting, and violent. This chapter considers both conflicts in a single case study because together they comprise a somewhat uniform body of British light infantry experience in low-intensity conflict. Comparing and contrasting these campaigns provides a useful analysis of the nature of light infantry and light infantry combat.

In Malaya, the Communist insurgency had its origins in the organizations established by the Malayan Communist Party (MCP) during World War II to fight the Japanese. Trained, armed, and supplied by the British, the military arm of the MCP—known as the Malayan People's Anti-Japanese Army (MPAJA)—grew into an extensive and efficient organization, some elements of which operated under British liaison officers. After the war ended, the MPAJA was disbanded and supposedly disarmed. However, the hard-core Communist elements of this small army hid their arms and supplies in secret caches for future use.

For a time, the MCP cooperated with the reestablished colonial administration in Malaya. When it became clear, however, that the aims of the MCP to influence the establishment of a socialist-type "People's Government" had no chance for success, the MCP adopted a more violent policy of social destabilization through labor unrest, strikes, and eventually, armed uprisings and acts of terrorism. In implementing this policy, "Vast quantities of rubber were stolen, rubber estate offices were burned down, British planters and miners and their Chinese, Indian and Malay employees were murdered" (see map 10).[1] In response, the government of the Federation of Malaya (hereafter referred to as the Federation) declared a state of emergency on 18 June 1948 and adopted emergency powers to deal with the violence. In addition, the MCP was outlawed on 23 July 1948.

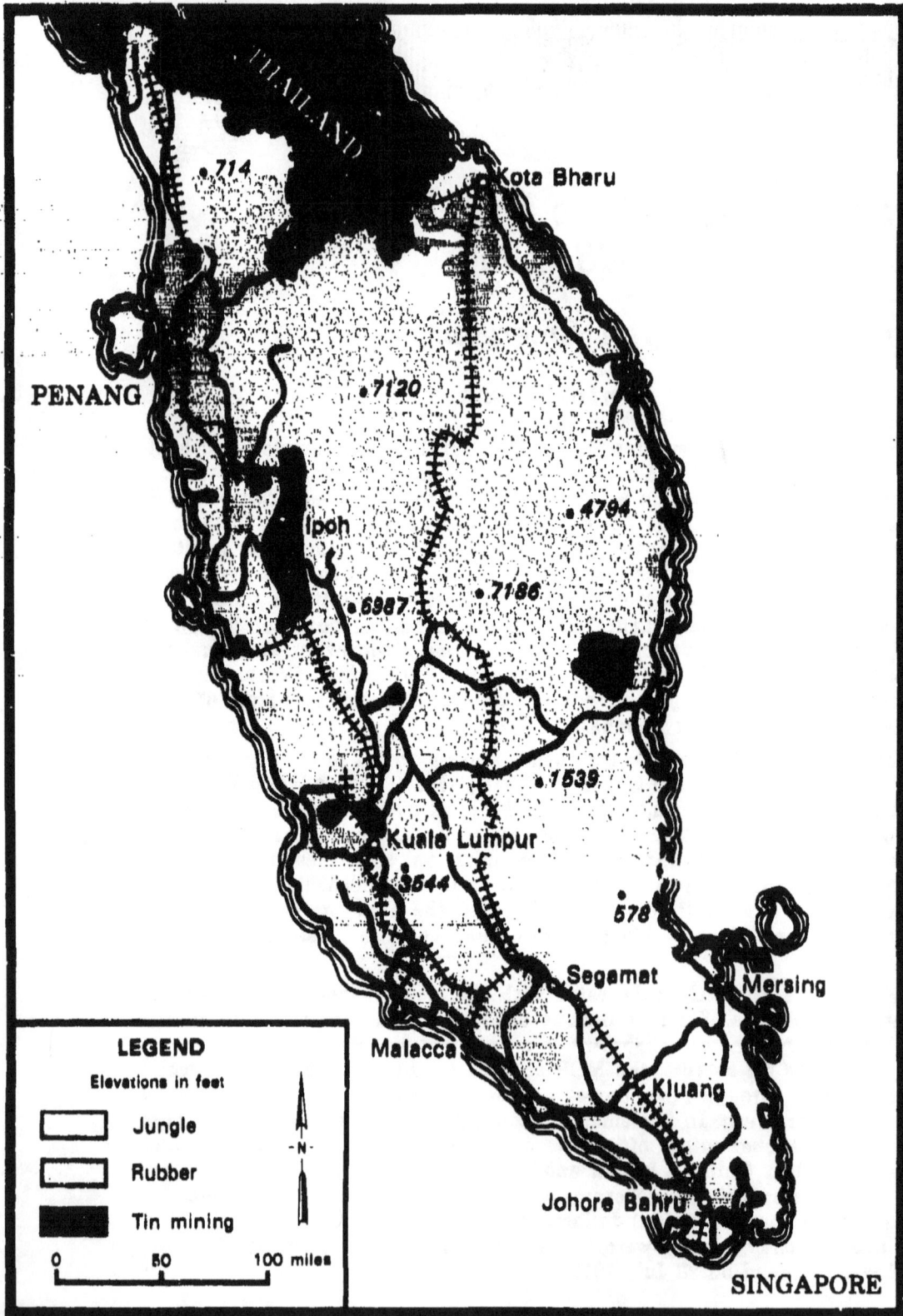

THAILAND

.714

Kota Bharu

PENANG

.7120

.4794

Ipoh

.6987

.7186

.1539

Kuala Lumpur

.3544

.578

Segamat

Mersing

Malacca

Kluang

LEGEND

Elevations in feet

- N -

Jungle

Rubber

Tin mining

0   50   100 miles

Johore Bahru

SINGAPORE

Source: Miers, *Shoot to Kill*, 217

Map 10. Malaya

Initially, the MCP enjoyed significant success. Even though its armed elements, now known as the Malayan Races Liberation Army (MRLA), never grew beyond a strength of 8,000 men and women, the MCP held sway over a large portion of the population and brought the huge rubber industry to a standstill.[2] Caught short by the suddenness and scope of the outbreak, the British introduced major developments that enabled them to regain control of the situation. These developments were the adoption of the Briggs Plan and the establishment of the Jungle Warfare School (JWS).

Named after its primary author, General Sir Harold Briggs, and adopted on 1 June 1950, the Briggs Plan aimed to bring the population of Malaya under closer administrative control and to isolate the guerrillas. Based in part on a study done by Michael Calvert, the former Chindit 77th Brigade commander, the plan had several main features. First, it required the rapid resettlement of isolated squatters and villagers into areas under the surveillance of the police, Home Guards (a paramilitary defense force), and the army. It also called for the consolidation of the local labor in mines and on estates to provide this rural population with more security. Furthermore, it instituted a thorough program of food control to deny material support to the MCP and MRLA. In addition, it called for a strengthening of the intelligence network through recruitment and training of criminal investigators and Special Branch police personnel (intelligence). Finally, it established a joint framework for coordinated activities between the civil, police, and army organizations. The Briggs Plan acknowledged that the conflict would be protracted and laid the foundations for a long-range solution.[3] Ultimately, over 600,000 villagers were resettled under the plan.[4]

The Briggs Plan embodied the overall strategy for solving the problem of the Emergency. The JWS provided the doctrinal basis and training for the tactical operations by army forces against the guerrillas. Established in 1948 at the Far East Training Center in Johore Bahru, the school was organized by Lieutenant Colonel Walter Walker, a three-year veteran of the Burma campaign in World War II. Basically, the school ran a six-week course for unit cadres and a six-week course, primarily cadre taught, for unit main bodies. Training included instruction and exercises in land navigation, marksmanship, quick fire, patrolling, jungle tactics, ambushes, tracking, and the use of jungle resources. Graduation exercises were live patrols in areas where guerrillas were known to be operating. Every battalion deployed to Malaya passed through the JWS before being committed to actual operations. The improvement in tactical operations by battalions trained at the JWS forced the MRLA to call off its large-scale operations and form into smaller, hit-and-run units known as Independent Platoons.[5]

By the end of 1951, the Briggs Plan and the improved tactical performance of the security forces impelled the MCP and MRLA to give up the initiative, turning from the role of the hunter to that of the hunted. The appointment in January 1952 of General Sir Gerald Templer as high commissioner hastened this transition. Through Templer's forceful leadership and insistence on full cooperation between the various civil, military, and police agencies, the guerrillas were driven deeper into the jungle. The fighting of the war and the

Table 4. State and District War Executive Committees

SWEC	DWEC	Responsibilities
CIVIL		
State prime minister (PM)	District officer	Local government
Executive secretary		Affairs of PM's office
Information officer	Information officer	Public relations/PSYOP
POLICE		
Chief police officer	Police commander	All police in area
Head, Special Branch	Special Branch officer	Police intelligence
Home Guard officer	Home Guard officer	Home Guard units
MILITARY		
Brigade commander	Battalion commander	All military troops in area
Military intelligence officer	Military intelligence officer	Military intelligence

Note: Community leaders, local planters and industrialists, and invited specialists occasionally attended meetings.

civil administration of the country, Templer stated, were "completely and utterly interrelated." It was during his tenure that the insurrection was essentially brought under control and the patterns of antiguerrilla operations established.

The Briggs Plan directed a joint civil-military-police approach to eliminating the insurgency. The plan focused on the Federation of Malaya, which included nine Malay states and two Straits Settlements, each of which was further divided into circles or districts. At each echelon, a War Executive Committee (WEC) was established that was headed by a senior political administrator (chief minister or district officer) and which included the senior police officer, military commander, Home Guard officer, and information officer in the area, plus others as required (see table 4). These state and district WECs carried out the policies established from the federal level by the high commissioner and director of operations, which were establishment of curfews, food control, route control, and direct operations. This kind of integration was essential to ensure that the security forces acted in support of the government and that the independent chains of command did not function at cross-purposes.[6]

Following decisions taken during State and District War Executive Committee (SWEC and DWEC) meetings, military operations were planned and supervised from the Joint Operations Rooms (JOR), established most often in police headquarters at each level. Infantry brigade or battalion intelligence officers ran the JOR and remained abreast of conditions at all times. Essentially, the JOR functioned as a clearing house for intelligence from the police, military, and Special Branch. One of the most important functions of the JOR was to provide clearances to unit patrols to operate with a relatively free rein in certain areas thought to harbor guerrillas. Most battalions conducted regular, daily meetings with police personnel at the JOR. Lasting an hour or so, these meetings—"morning prayers," as they were called—enabled each service to review new intelligence and to be informed of the progress of ongoing and near-term operations. The JOR also maintained radio and telephone links with subordinate police and military units.

At the peak of the antiguerrilla war, the Malayan government employed 40,000 soldiers, 45,000 police, and 1.25 million Home Guards to root out the insurgents and protect the population.[7] As mentioned earlier, the Communist terrorists numbered only about 8,000 at their peak. Such a large imbalance of force, though, is not unusual in insurgent warfare.

As part of the Emergency, the British employed all kinds of infantry units. Of the more than thirty-five battalions deployed to Malaya from 1948 to 1960, the largest number came from Gurkha regiments stationed in the Far East. Australia and New Zealand also provided battalions. However, a significant number of regular and motorized-mechanized infantry units from Great Britain and other areas were also used in the Emergency, including, for example, the Green Howards, the Gordon Highlanders, the South Wales Borderers, the Somersets, and others. Although almost all of the regular infantry battalions (and some of the Gurkhas) had little or no jungle light infantry experience, the JWS course provided them with the skills and attitudes required to operate against the guerrillas until actual operations refined their expertise. Interestingly, many of the soldiers used in Malaya were first-term drafted soldiers on short tours. Nonetheless, these soldiers proved more than capable, although they did complicate the training problem and cause unit turmoil through their high turnover rate.[8]

The Threat

The Malaya Communist Party (MCP) maintained an organization quite similar to that of the Federation (see figure 9). While their strength varied from one district to the next, each district contained at least one independent Communist platoon. These units communicated within and between districts primarily by couriers.

The Communist terrorists (guerrillas)—who were designated CTs by the British—were armed primarily with the small arms left over from the MPAJA. They received no significant external aid from China or the USSR. The main strengths of the CTs were their flexibility, discipline, hardihood, attitude, and ability to react quickly. Careful and wary, they demonstrated adept jungle craft in their contacts with security forces. As jungle fighters, they were worthy opponents.

After the arrival of General Templer, the CTs abandoned their policy of frequent confrontation. Instead, they now aimed at simply remaining in being, hoping to preserve their strength. Thus, they broke up their larger units and adopted evasive tactics. Avoiding direct clashes with the army or the police, small groups of guerrillas focused on careful ambushes, quick raids, and terrorism of the local population when they failed to cooperate.[9] As a result, the army had to root them out of the jungle singly and in small groups. By the end of 1952, progress and success at battalion level were measured in terms of the number of kills a unit produced.

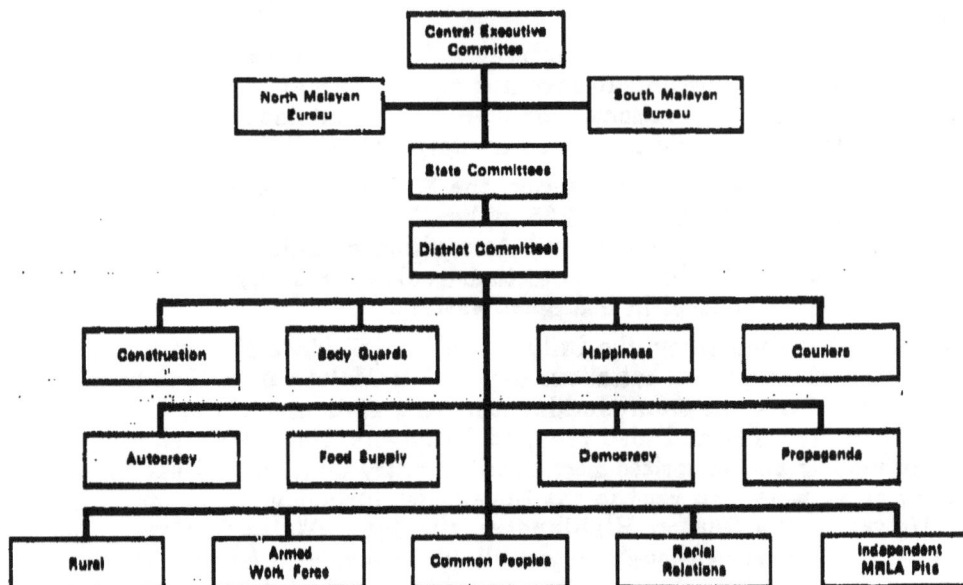

Figure 9. Organization of the MCP and MRLA

The great weakness of the terrorists was their reliance on civilian support for food and information. Denial of such support through the provisions of the Briggs Plan was an important tenet in the British operational concept. Implementing the Brigg's Plan, however, was difficult given the number of Communist sympathizers (estimated at 60,000 by one source) and the ability of the CTs to maintain contact with the population.

Operational Concepts

The primary theme of British military operations in Malaya was the painstaking, long-term, and systematic elimination of Communist terrorists from the entire country. This process required classifying the various districts in the country by color. Terming a district "black" meant that the CTs retained significant capabilities there. A "white" district was an area that had been cleared of CTs to the point that tactical operations were no longer required there, and all or most of the civil rights suspended during the Emergency could be restored. But not all areas could be cleared at once. Thus, security forces in one district might be required to maintain a holding action against the CTs, while civil and military efforts were applied in another area to wipe out the CTs and establish "white" status there. The British effort initially went to the areas where the Communists were weakest; then they cleared the "blackest" areas. Success of antiguerrilla operations could be measured in terms of how much of the Federation had been declared "white" and how fast it was changing from "black" to "white." Once an area was declared "white," it remained the responsibility of the police and the Home Guard to preserve its "white" status against the reemergence of the guerrilla organization. Patience, harmony, and cooperation were indispensable to this approach.

100

Active military operations assumed three main forms.[10] First, combat patrols moved out with specific objectives based on intelligence produced at the JOR or provided by Special Branch informers. Such actions varied in size and duration depending on the target; they were frequently unsuccessful. The second method, described by one Malayan commander as a "partridge drive," involved the saturation of an area with large units—battalion to brigade size—with the object of flushing out CTs by employing sheer numbers. Sometimes, these operations lasted two to three months. However, they were even more unsuccessful than the first method, because they were difficult to disguise. The CTs were seldom taken by surprise, and the jungle offered too many places for them to hide from the bushbeaters. Battalion and larger operations based on more or less conventional tactics just did not work well at all.[11] A more subtle means was required.

The third operational method, sometimes called jungle bashing, made the most of the highly developed jungle craft of the British infantry and also exploited the best available intelligence on the enemy. Using this approach, units quietly deployed several patrols from squad to platoon strength into an area where the guerrillas were known or thought to be. Each platoon established a temporary base (24—48 hours) from which it pushed out smaller patrols in a systematic fashion to cover thoroughly and carefully a designated area. Once one area had been checked out, the platoon moved on to a new area, and the process was repeated. In this manner—while being resupplied, if needed, by air, road, or cache—the units could thoroughly investigate a jungle area in two or three weeks, while maintaining secrecy. This saturation patrolling frequently produced contacts; the British were often able to surprise the terrorists in their jungle camps or on the march.

In the last half of the Emergency, company-size bases were established on a thirty-day basis. These bases were fortified to an extent, although they were not especially vulnerable to attacks by the weak CT units then in being. The bases functioned as the administrative, logistical, and command and control centers for the platoons and squad patrols emanating from them. After a base was maintained for a month, it was closed down and the company removed. A new company was then moved into a new area. The British used this technique whenever they wanted to force a group of CTs to abandon an area. Forced to move by this pressure, the Communists were vulnerable and subject to ambush. Even if no contacts were made, this method caused the CTs to lose their support and information in one area and to begin anew in another.[12]

The operational concept described above required that certain principles be observed. One of these was that close coordination of civil, military, and police actions had to be maintained. Equally important was the need to sustain decentralized, offensive, and extended operations that granted junior commanders a wide latitude in decision making. Decentralization was necessary in operations because of the wide area to be covered, the limited forces available, the dispersion of the CTs, and the demonstrated failure of large-unit operations. Since fights took place almost exclusively at the team, squad, and platoon level, the commanders of units needed to have a free hand to exercise their own judgment in the field. Company commanders, in particular, had to be given broad discretion so that they could independently plan and execute their intentions based on their own assessment of the situations in their areas. This was especially necessary because higher-level commanders lacked both the physical means and the inclination to visit their units daily or to approve every tactical plan.

Because the army units were hunting small groups of CTs—sometimes even single individuals—operations were necessarily extended in time. Once a patrol entered the jungle, it normally was prepared to stay out for four to ten days. In some cases, units stayed in the jungle without relief for as much as thirty days. Units retained a pronounced offensive frame of mind, ready to spring into action at the first sight of the enemy.

Patrolling a coconut plantation

Applying all of these operational methods permitted the British to maintain unrelenting pressure on the guerrillas. Constant harassment kept the enemy on the move, disrupted his command and control, confused him, frayed his nerves, and prevented him from carrying out his operations.[13] When this kind of pressure was combined with police-directed curfews, searches, psychological warfare, and stringent food denial programs, the enemy frequently surrendered out of hunger and a broken spirit.

Intelligence in Malaya

"There is no doubt that the soundest (and, in the end, the cheapest) investment against Communist insurgency in any country is in a strong, handpicked, and well-paid police intelligence organization, backed up by the funds to offer good rewards."[14] The above quotation typifies commentaries on the Malayan Emergency, most of which agree that timely, accurate intelligence was the most important ingredient leading to tactical success against the terrorists. But in the early stages of the conflict, obtaining accurate information was a difficult problem. Villagers, too terrified by the terrorists to cooperate with authorities, provided only a few contacts and little information.[15] As the guerrillas suffered casualties and retreated into the jungle, however, the intelligence flow increased in volume and reliability. Nonetheless, the need for intelligence became even more acute as the CTs adopted more evasive policies. The problem for the British forces after 1951 was one of putting military forces into contact with the insurgents. Success increasingly depended on reliable intelligence.

The ability of the army to produce intelligence on the locations and movements of the CTs was limited. The best commanders spent "long hours in tactful discussions with police officers, administrators, rubber planters, tin miners, and local community leaders, getting them to cooperate with the soldiers and to promote the flow of information to them."[16] To a large degree, the army simply had to rely on the police intelligence organization, the Special Branch, and other civil agencies for information. Of these organizations, the Special Branch, by far, developed the most intelligence. The joint organization of civil, military, and police agencies established to prosecute the war lent itself well to a comprehensive exchange of information between the services.

The Special Branch, for its part, painstakingly built up the enemy order of battle based on several sources: CT food suppliers, captured documents, informers, surrendered and captured personnel, personal reconnaissance, and its own intimate knowledge of the area. Most of the Special Branch officers were unusual men—energetic, insightful, extremely dedicated, and well suited to the physical and intellectual demands of their positions. Through the use of impressive cash awards, mild (legal) coercion, and the promise of immunity, the Special Branch lured many Communist sympathizers to betray their former comrades. Though the information acquired by the Special Branch often took a long time to mature, the support it provided the British was priceless. Army commanders unanimously praised the Special Branch for its cooperation and competence.[17]

Tactical success and the flow of public information were inextricably linked. Thus, the more guerrillas that were killed by the infantry, the more information came in. This increase in information subsequently led to even more kills, and so on. The pump always had to be primed, however, by the confidence of the population that the security forces (particularly the police posts) could protect them.

One of the best sources of information for the Special Branch and the army was surrendered enemy personnel. Disillusioned by their leadership, despondent about their cause, harassed constantly by British patrols, many CTs proved vulnerable to the promise of lenient treatment should they decide to surrender. It was well known, for example, that few surrendered personnel had ever been prosecuted for their former actions. So, many came forward when the opportunity arose.

The British used these surrendered men against the CTs in many ways. Frequently, they led British patrols straight back to CT jungle camps within twenty-four hours of their surrenders. Some were armed and used as guides and trackers for long periods of time. Others composed psychological operations tapes that were then broadcast by aircraft loudspeakers to induce other surrenders. Surrendered Communists invariably provided oral information or documents concerning the personalities, plans, and methods of the MCP and MRLA. A great deal of the information maintained by the Special Branch concerning the Communist terrorists' order of battle came from surrendered enemy personnel. No surrenders, however, occurred without tactical military pressure.

Aerial reconnaissance and extensive patrolling also produced useful intelligence regarding the existence of camps, food dumps, jungle gardens, trails, and terrain. This kind of information was then used as the basis for directing future patrols. When closely analyzed, this intelligence produced reliable predictions on likely enemy locations and movements and increased the probability that contacts with the enemy would be made.[18]

Jungle Tactics Against the Malayan Insurgents

The central, omnipresent task of the British light infantry in Malaya was to go into the jungle, find the enemy, and kill him by surprise as often and as quickly as possible. This task, coupled with the nature of the enemy and the environment, led directly to the adoption of the principles described earlier—decentralization, offensiveness, extended operations, relentless military pressure, and the granting of wide latitude to junior leaders and commanders. In implementing these principles, however, the infantry did not wait for the CTs to act; they doggedly and expertly hunted them down in their jungle hideouts and ambushed them at trails and contact points.

The specific tactics employed by the infantry are described in lucid detail in a pamphlet entitled *The Conduct of Anti-Terrorist Operations in Malaya.* Produced by the Jungle Warfare School and widely known as the "Atom manual" (see bibliography and appendix A for further information), this pamphlet functioned as a tactical bible. Three editions were published from 1952 to 1957; each incorporated lessons learned through actual combat experiences. Space prohibits a complete discussion of all of the techniques and SOPs contained in the manual; however, five main areas--tactical organization and equipment, field craft, patrolling, ambushes, and attacks—are outlined below.

Tactical Organization and Equipment

The ATOM manual acknowledged the failure of conventional tactics and organization when applied to counterinsurgency warfare. Instead, it prescribed the need for flexibility based on the factors of METT-T (mission, enemy, terrain, troops available, and time). As a result, no standard platoon and company organizations were used in Malaya. The organization adopted was made to fit the situation but was always light and mobile.[19]

Squad organization also varied, although it most often assumed a 3x3 grouping reminiscent of the CCF in Korea. Each squad consisted of a reconnaissance group or point group, a support group manning a light machine gun, and a rifle group of three riflemen. The ATOM manual strongly advocated the 3x3 organization for several reasons:

- It simplified the squad leader's job of control.

- It provided the grouping needed for the effective minor tactics that had been evolved for use against the CT.

- It helped to train the potential junior leaders who could take over a section if necessary.

- It provided small three-man teams, which experience had shown to be good basic teams.[20]

Within these squads, care was taken to keep rucksack loads as light as possible. For example, only one set of spare clothing was taken so that soldiers could sleep dry. Wet, dirty clothing was redonned each morning. Moreover, the use of underwear was not recommended on the grounds that it could lead to skin infections. Hammocks were used for quick, dry sleeping perches. The men also carried improvised canvas strips for shelters and stretchers (when needed).[21]

The primary infantry weapon was the 7.62-mm self-loading rifle, although carbines and submachine guns were also issued. Reconnaissance and point groups carried shotguns for punch and lethality at short range. Soldiers also used shotguns on night ambushes for the same reason. The Bren gun (light machine gun) formed the main unit of firepower in the squad. Interestingly, the 2-inch mortar was deemed unsuited for operations in the Malayan jungle. Instead, soldiers used a smaller, short-range grenade discharger.[22]

Field Craft

The success of small-unit tactics depended entirely on the quality of field craft employed by units. Movement through the jungle required concentration and meticulous attention to detail to avoid being surprised or tipping off the CTs. Wary, alert, senses honed for survival, the terrorists quickly spotted any inadvertent signs of the British presence. To avoid their telltale scents, British troops used no chewing gum, tobacco, toothpaste, hair tonic, or insect spray. Even the use of soap and the taking of baths by soldiers was taboo while they were on patrol or maintaining an ambush site. To limit the chances of detection, leaders also controlled fires, cooking, and eating. Furthermore, prior to leaving an overnight patrol base, soldiers erased every trace of their presence.

Trudging through Malayan swamps

For soldiers to detect the passage or presence of the guerrillas required finely tuned powers of observation. All troops received training in stalking techniques and in spotting significant jungle signs, such as overturned leaves, bent twigs, bruised blades of grass, or pieces of bark cut by passing humans. The examples set by aborigine trackers aided the development of such skills among the rank and file soldiers, who, in emulation, sometimes became as adept as the trackers themselves.

Expert field craft also demanded extremely high standards of discipline and patience as well as scrupulous attention to the environment. One miscue or careless act could put a patrol in jeopardy or send a quarry out of harm's way. Maintaining such high levels of field craft was largely a function of good training at the Jungle Warfare School and in unit areas and of good, strict leadership while in the jungle. The infantry needed little prodding from their leaders to stay alert, however, because they respected the skills of their opponents. Moreover, they knew one of the primary lessons of jungle combat: In jungle warfare, where most contact comes as a surprise to both sides, the man who shoots first survives.[23]

Patrolling

Infantry units in Malaya spent more time patrolling than in any other type of tactical activity. All patrols moved out with a clear mission, defined by the company commander in most cases. Activities on patrol were strictly regulated by adherence to detailed SOPs, by high standards of field craft, and by good jungle sense. Great patience and concentration were required, because hundreds of hours of patrolling were needed to make contact with the enemy.[24] Patrols often stayed in the jungle for one to two weeks without seeing a single terrorist.[25]

The normal rate of movement in the jungle ranged from one-half mile (average) to one mile an hour. Land navigation was difficult; dead reckoning—that is following a compass bearing and measuring one's pace—appeared to be the most common means. Cutting the bush was avoided; instead, patrols pushed carefully and steadily through it creating little noise. The principle that one must "never fight the jungle" dictated this methodical, patient approach.

Because men who are still have the tactical advantage in the jungle, patrols were most vulnerable to enemy ambush or detection during movement. Patrols learned to pause and listen for ten minutes for every ten or twenty minutes of movement. Patrols also moved during heavy rainfall to shield the noise of their movement. Conversation among the men was restricted to low whispers.

Two basic formations were used. In thick jungle, the squad patrol moved in single file. In clearer terrain, the patrol adopted an open formation (see figure 10). In each case, the squad leader (and a guide, if one was used) followed directly behind the reconnaissance group. Captured terrorists, surrendered enemy personnel, and local inhabitants—anyone who had intimate knowledge of the area—functioned frequently as patrol guides.[26] Some surrendered enemy personnel established long relationships with units and coached the soldiers on proper field craft.

When following the terrorists' trails, patrol leaders often employed Iban or Dyak trackers, who were brought over from Sarowak in North Borneo as early as 1948.[27] Artists in tracking and experts in jungle lore, these men became very much a part of the units they supported and were highly honored. Capable of following trails for days, these aboriginal trackers worked in pairs and walked at the heads of squads, with two other scouts following directly

SINGLE-FILE FORMATION

1. This formation will *not* be used in rubber or other plantations.

2. Single-file formation is used in the jungle where troops cannot move in a more open formation.

3. Distances between individuals and groups will vary according to visibility.

4. Generally there should never be less than five yards between each man. Distance between groups should be governed by the nature of the ground and vegetation and the necessity for maintaining control.

5. A tracker group, if accompanying the patrol, will be located in a suitable position for immediate employment.

OPEN FORMATION

1. "One Up":

  (a) *Advantages*.

    (1) Ease of control.
    (2) Good fire power to front and flanks.
    (3) On contact, the leading group only is committed and two are available to maneuver.

  (b) *Disadvantages*.

    (1) With the fleeting targets that are offered in Malaya, fewer men are likely to see the CT on first contact.

2. "Two Up":

  (a) *Advantages*.

    (1) A wider front is covered.
    (2) The formation is less vulnerable to ambush.
    (3) More weapons are available to fire forward in event of a sudden contact.

  (b) *Disadvantages*.

    (1) On contact, the two forward groups may be committed and there are less troops available for maneuver.

3. "Three Up": Although this formation will cover more frontage, it is difficult to control and allows nothing for maneuver.

4. Distances between individuals and groups will vary according to the ground through which troops are passing.

LEGEND

S1—Leading Scout
S2—No. 2 Scout
SC—Recce Group Commander
PC—Section or Patrol Commander
G—Guide
B1—Bren Gun No. 1

B2—Bren Gun No. 2
BC—Support Group Commander and Section 2 IC
RC—Rifle Group Commander
R1—No 1 Rifleman
R2—No. 2 Rifleman

Source *The Conduct of Anti Terrorist Operations, Malaya* (Third Edition, 1958) II 6

Figure 10. Patrol formations

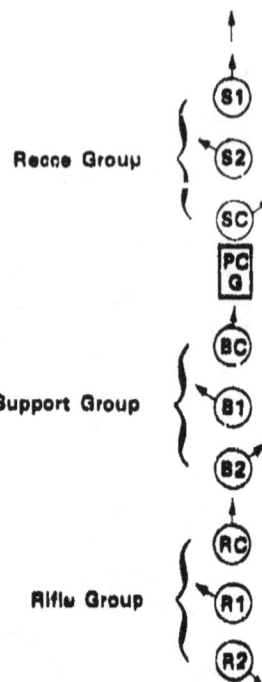

OPEN FORMATION—ONE UP

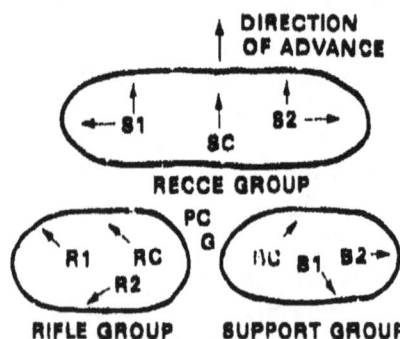

OPEN FORMATION—TWO UP

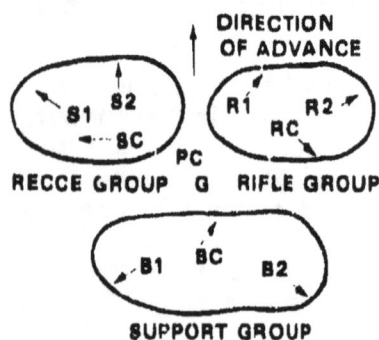

Note: Small arrows indicate the direction of responsibility for observation

behind them, whose sole job was to protect the trackers. Tracker dogs proved to be much less reliable. They often left the trail for the smell of water.[28] Tracker dogs required handlers, too, and a protective scout team.

Patrols normally stayed in the jungle for four to ten days, although two- to four-week operations were not uncommon, particularly if ambushes were the object. In some cases, extreme distances were covered.[29] The most effective method of operation was the establishment of a temporary platoon base, from which circulated a number of small patrol elements in varying directions. Systematically exploring a given area, these small teams moved out without rucksacks or rations. If an occupied enemy camp was found, one member of the patrol went back to the base to inform the platoon leader, who then planned and led a platoon-size raid on the camp as quickly as possible.

The size of these circulating patrol elements depended on the courage, confidence, and field craft of the patrols' members. In some cases, two-man teams were used, in other instances, whole squads. Whatever the size, these elements took care to avoid bumping into each other.

No terrain was avoided. The patrols hunted the guerrillas wherever they holed up, including swamps. Sleeping dry in hammocks above the water, swamp patrols spent the days in thigh- to chest-deep water.

Patrols generally did not move at night because of the poor visibility in the jungle. Instead, they occupied a small, temporary patrol base as illustrated in figures 11 and 12. On the other hand, patrols did move at night through rubber plantations because of the cleared undergrowth and good visibility.

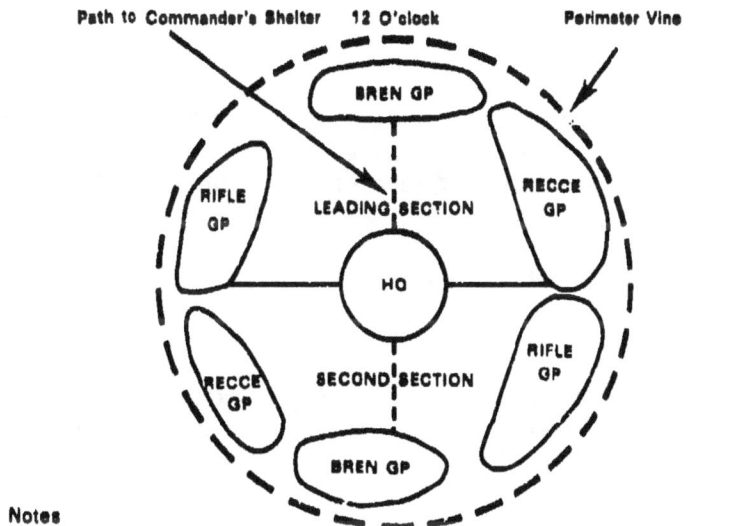

Notes

1. The patrol commander having indicated 12 o'clock, the leading section of the patrol moves into position between 0300 hrs., 1200 hrs., and 0900 hrs. The second section moves into position between 0300 hrs., 0600 hrs., and 0900 hrs.

2. If the layout is kept standard within subunits, every man and group will know their approximate positions and who will be on their left and right.

Source: The Conduct of Anti-Terrorist Operations. Malaya (Third Edition. 1958). II 8

Figure 11. Suggested layout of a two-squad base

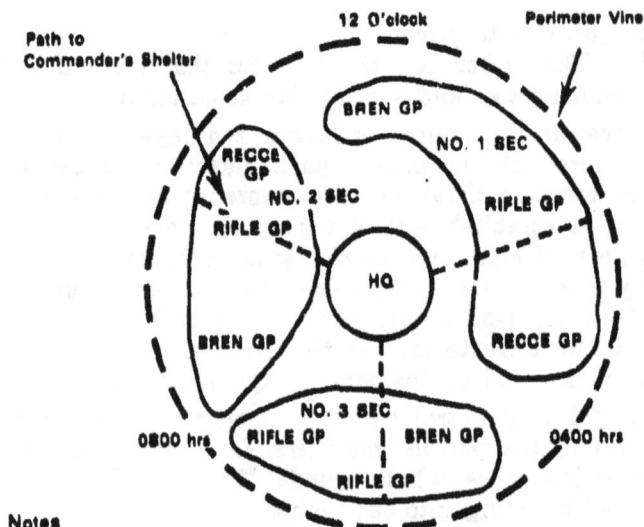

Notes

1. The patrol commander having indicated 12 o'clock, No. 1 Section moves to take up position between 12 o'clock and 0400 hrs., No. 2 Section between 12 o'clock and 0800 hrs., No. 3 Section between 0400 hrs and 0800 hrs.

2. The entrance to this base is at 12 o'clock.

Source: *The Conduct of Anti-Terrorist Operations, Malaya* (Third Edition, 1958), II-6.

Figure 12. Suggested layout of a three-squad base

When leaving the jungle to return to their garrisons, patrols usually exited at a certain point and specified time to avoid being mistaken for guerrillas themselves. Moreover, company commanders always kept a patrol on standby in readiness to react to a guerrilla raid or a timely tip on guerrilla locations. Finally, patrols never used reconnaissance by fire in Malaya in any fashion. (For further information, a patrol aide-memoire from the ATOM manual is reproduced in this chapter at appendix B.)

Ambushes

Whenever the flow of intelligence promised the likelihood of an enemy contact, the light infantry dispatched ambush patrols to lay in wait. Commanders often formed special ambush teams from men noted for their marksmanship, field craft, or other particular quality (such as familiarity with the area of the objective).[30] A high proportion of these ambushes occurred on the jungle fringe where the terrorists met with their food and information sources. The British established strict rules in regard to the conduct of such ambushes and spent a good deal of garrison time training for them. Success depended on adequate preparation, which included a thorough plan, weapons check, equipment check, briefing to all members, and a rehearsal. Night ambushes were rehearsed at night. Because each ambush was considered unique, the patrol organized for the ambush varied in size and content based on the situation.

The ATOM manual cites a number of conditions that are essential to the performance of successful ambushes:

- Good shooting from all positions—kneeling, sitting, standing, lying, and firing behind cover.
  - A high standard of training in ambush techniques.
  - Careful planning and briefing.
  - First-class security in all stages of the ambush.
  - Intelligent layout and siting.
  - Concealment.
  - A high standard of battle discipline throughout the operation.
  - Determination by all members of the ambush party to wait and kill.
  - A simple clear-cut plan for springing the ambush.[31]

Of these principles, perhaps the most important was the requirement for strict, unflinching battle discipline. These principles are explained in further detail in the manual.

Although 80 percent of the British ambushes were sprung within nine hours of the occupation of a site, the infantry also mounted some long-term ambushes. Naturally, these operations were more complex, requiring the establishment of a base area some distance away from the ambush site. Arrangements for feeding, sleeping, security, and relief of the ambush group also had to be made. Ideally, the patrol contained three separate groups: one at the ambush site, one at rest, and one in reserve. If the patrol was too small for this organization, procedure demanded that the ambush group simply retire from the ambush site when it was necessary to eat and sleep, and to return later.

In this manner, the light infantry held themselves in position to ambush for some incredibly long periods of time. One platoon of the Green Howards staked out the house of a terrorist food supplier for twenty consecutive nights.[32] In another case, a patrol maintained an ambush for ten days and nights. Given seven days' rations, they simply were told to make them last ten days.[33] In one of the longest cases on record, Brigadier Walter Walker, commanding the 99th Gurkha Infantry Brigade, left a patrol in place for twenty-seven days. On the twenty-eighth day, the terrorists finally entered the killing ground and suffered two dead and one surrendered.[34] The results of such long-term operations may seem paltry compared to the sacrifice made, but these methods were the only guarantee of success. To succeed, ambush groups had to achieve high standards of field discipline.

To lessen the chance of ambush failures, the ATOM manual analyzes the reasons for failures and suggests remedies:

- Disclosure of the ambush by the noise made by cocking weapons and moving safety catches or change levers. Check your weapons, practice men in their silent handling, and ensure that all weapons are ready to fire.

- There was a tendency to shoot high at the light face of the terrorist. This must be corrected on the jungle range.

A native tracker from Sarawak

- Disclosure of the ambush position by footprints made by the ambush party moving into position and by movement of individuals at the crucial time, when the CT were approaching.

- There was a lack of fire control, and commanders were unable to stop the firing and start the immediate follow up.

- Commanders were badly sited with consequent lack of control.

- There was a lack of all-round observation resulting in CT arriving in the area of an ambush unannounced.

- There were misfires and stoppages through failure to clean, inspect and test weapons and magazine.

- There was a lack of a clearly defined drill for opening fire, and orders were contradictory.

- There was a tendency for all to select and fire at the same target.

- Fire was opened prematurely.[35]

112

The Attack

The typical targets of attacks by British light infantry in Malaya were small parties of guerrillas hiding out in jungle camps, who were discovered by patrol elements. Attacks against these guerrillas took several forms. One type of attack took place when the guerrilla sentry recognized the approach of the infantry patrol and took it under fire. In this situation, the patrol executed an immediate attack drill by sweeping through the camp as quickly as possible. This kind of attack seldom succeeded in killing many terrorists, because they deserted the camp immediately after hearing the first shot.

Another type of attack—the deliberate attack—was mounted when the terrorist camp was discovered while the presence of the British unit was still concealed. The patrol element that discovered the CTs kept the camp under surveillance and sent off a runner to inform the platoon leader or company commander, who quickly reached a decision regarding the time and form of action. If time (and skill) permitted it, the attack leader conducted his own reconnaissance of the site, being careful not to alert the enemy. By virtue of previous training and rehearsals in garrison, the attack leader was able to limit his attack orders to a few fundamental points, the others being covered by SOPs.

The attack force comprised two elements—an assault party and a cutoff party. Both elements secretly approached the camp, crawling the last 100 meters or so to their final positions. Troops often were able to crawl in this way to within five to twenty meters of the enemy perimeter. The crucial maneuver in the attack was the encircling of the CT camp. It had to be accomplished with great patience and stealth. Unless the camp was completely surrounded, the majority of terrorists usually escaped.

The assault began after sufficient time had passed to permit all elements to reach their positions. As soon as the assault party attacked, sometimes from two directions, the men in the cutoff party assumed the best possible firing positions and waited for targets to appear. SOPs demanded no indiscriminate firing and no movement out of position. Usually, there was a signal of some sort for a cease-fire. If any of the terrorists escaped, the patrol followed up on their trail as soon as possible after reporting the results of the raid, collecting material with intelligence value, and treating the wounded.

Speed was crucial in these operations. Jungle camps discovered in the evening might be deserted by daybreak. The guerrillas were adept at collecting their weapons and gear and disappearing in seconds. Frequently, only a few hours passed between discovery of a camp and the execution of an attack. Early morning attacks seemed to be favored in order to cover the approach of the attacking force, to catch the enemy while he was rising, and to leave as many hours of daylight as possible for pursuit, if required.

Combat Support

For combat support, the British Army in Malaya employed light armor, artillery, helicopters, and air support. No engineers were used, although one participant in the Emergency, Brigadier M. C. A. Henniker, believed that they could have been put to good use in improving and constructing roads, trails.

and jungle bases and in building bridges over streams.[36] Light-armored cars were used to escort motorized convoys and to patrol routes capable of being interdicted by the guerrillas.

Infantry commanders used artillery bombardments to harass the guerrillas and to keep them on the move. Artillery units might, for example, fire on suspected enemy camps or previously abandoned camps to discourage their reuse and to force the terrorists to leave them if they were occupied. Patrols then tried to pick up enemy tracks or listen at selected points for enemy movement. Field artillery was also used to soften up the enemy preparatory to a psychological operations campaign to induce surrenders.[37] To a great degree, however, artillery fires proved to be quite unproductive. Although the British fired thousands of shells into the jungle, they rarely obtained hits or kills as a result. Moreover, there was also the danger in the deep jungle that innocent aboriginal tribesmen might be hurt.

The limited capabilities of the early models of helicopters and the absence of experience in their use prevented comprehensive employment of helicopters in support of the infantry. Helicopters, however, were used for casualty evacuation (as were light aircraft), tactical deployment of small units, exchange of police garrisons in remote locations, transport of unit commanders from place to place, aerial spraying of the guerrillas' jungle crops, and limited air resupply. But helicopters had no direct offensive role in Malaya.

Fixed-wing aircraft, in contrast, performed a variety of offensive and support tasks. Offensively, fighters and bombers conducted occasional air strikes against enemy camps. However, experience showed that these camps were hard to hit because they were rarely visible from the air. The British used bombers, nonetheless, to harass the guerrillas, to maintain pressure on them, to force them to move, to lower their morale, and to deny certain areas for their use or passage. In large, brigade-size "blitz" operations, like Operation Termite in 1954, bombers were used to seal off areas not covered by patrols.[38] However, the bombers had many drawbacks. They often missed, and they could hit one's own troops or the aborigines.[39] The infantry rarely, if ever, used aircraft for close-air support.

Another valuable service provided by aircraft was photo reconnaissance. Many a jungle crop discovered by air reconnaissance was later destroyed by aerial spraying or bombing. In addition, the air forces provided communications flights to provide relays and to determine and report patrol locations with their navigational aids. Furthermore, cargo aircraft delivered supplies by parachute to remote outposts and long-range patrols. Occasionally, they delivered to jungle airstrips. The British also conducted a few parachute operations for jungle rescue and to deliver Special Air Service (SAS) teams into the deep jungle for extended reconnaissance.

Finally, "voice" aircraft and leaflet drops embodied the two major techniques of psychological operations against the terrorists. Using tapes made by surrendered enemy personnel or other Chinese speakers with insight into terrorist psychology, these voice aircraft broadcast generous terms for surrender and advised the terrorists on surrender procedures. Truck-mounted loud speakers were used in the same fashion along the jungle fringe. Combined with stringent food denial and relentless military pressure, these operations frequently bore fruit.

114

Individual Skills

The quality of British and Gurkha light infantrymen was an essential ingredient in the defeat of the Chinese insurgents in Malaya. The British recognized that their troops had to be able to meet the terrorists man to man in the jungle and beat them. To accomplish this task, the army developed soldiers with highly refined light infantry skills.

Roping down from a helicopter—a rehearsal

115

The qualities required by light infantrymen included noiseless movement, powers of observation, intense concentration, rapid and accurate fire, fire discipline, land navigation, knowledge of the jungle, patience, fitness, ability to camouflage, and promptness in taking immediate action in accordance with SOPs. Perhaps the most important quality of these soldiers, however, was their mental attitude of self-discipline. The nature of the war in the jungle demanded that the British and Gurkha soldiers be ever watchful. Because one mistake might send the quarry vanishing into the jungle or expose the patrol to fire, soldiers had to be constantly alert and tuned in to the situation. Furthermore, they had to deny themselves many basic amenities enjoyed by regular infantry in conventional warfare: cigarettes, baths, shaves, hot meals, and conversations at a normal level of speech. In some cases, ambush patrols had to relieve themselves in place for days and go without food.[40] The mental attitude obtained by soldiers helped them to endure the nervous strain of operations. The mental stress on men during patrols often exceeded the physical strain.

Another light infantry skill that the British placed great value on was marksmanship. To develop this skill, company- and battalion-size garrisons frequently established their own ranges. There, soldiers trained on all the weapons assigned to the unit in addition to their personal weapons. Special attention was given to quick fire, during which soldiers stalked each other carrying air guns and wearing face masks in a controlled jungle terrain. Units tasked to mount night ambushes or night patrols usually first practiced on the ranges at night, since experience had shown that soldiers typically fired high at night. The high value the British placed on marksmanship is evident in all editions of the ATOM manual. All high commissioners-directors of operations in their short forewords emphasized the vital importance of quick, accurate shooting under all conditions.

When not operating in the jungle, infantry units often exercised police skills, assisting the local police in their duties, especially during major efforts of food denial or route control. Troops on this kind of duty functioned under police supervision, but in accordance with their own chains of command. Their duties included searching vehicles and individuals, checking identity cards, guarding detainees, manning road blocks and village exits, and maintaining order. Tedious and wearing, these duties required a different kind of patience and alertness. Unused to such close contact with the civilian population, the infantry had to practice a restraint that contrasted markedly with their jungle combat roles.

Leadership

The highest quality of leadership was necessary in Malaya if the light infantry operations there were to be successful. Providing this leadership were Briggs, Templer, and their successors, who, while providing strategic direction, established and maintained the framework for effective tactical operations at company level and below. The army's senior tactical leaders also exercised outstanding judgment by allowing the noncommissioned officers (NCOs), platoon leaders, and company commanders to conduct their activities without interference. Characteristically, the senior leaders remained willing to listen to the men on the ground. Thus, rather than impose their own ideas on the tactical units, they identified the best ideas in the field and distributed them throughout the army.

Just as the nature of the war brought out the best in soldiers, so it enhanced the performance of the leadership. Indeed, leaders provided such good examples in their positive guidance that it was reflected in the performance of their men. In addition, the leadership exhibited confidence, self-reliance, and the same high level of field craft as their men. Although decentralization in operations placed a heavier burden on leaders and increased their responsibilities, they rose to the challenge and displayed the necessary ability, imagination, and flexibility to react to unforeseen situations. This was important because a single decision by a patrol leader might send a squad on a three-day, forty-mile pursuit after fleeing guerrillas. The need to respond quickly and decisively was crucial to the effective conclusion of operations. Leaders had to possess a singleness of purpose and a relentless cast of mind to maintain the tactical efficiency and discipline of their men over the extended duration of each patrol and the campaign as a whole. When an opportunity to attack the guerrillas appeared, these NCOs and junior officers had to be ready to lead their men in violent, rapid action, often under very uncertain conditions. Decisiveness and an offensive spirit had no substitutes.

These traits were nurtured and developed both at the Jungle Warfare School and in actual operations, with the crucible of the field proving to be the best trainer, since success or failure usually depended on unit leadership. Decisiveness and the offensive spirit also grew in strength due to the decision by the higher leadership to place their trust, through decentralization, in their junior subordinates. Free of inhibiting higher interference, leaders were able to act aggressively.

Infantryman smeared with black grease paint

117

Logistics

Most logistical functions in Malaya were performed and based in the various garrisons. Lines of communication occasionally were interdicted by the terrorists, but they could be resecured relatively easily

In the early days of the war, air resupply of active operations was rare. Patrols carried the rations, spares, and ammunition that they needed on their backs. Because the patrols generally remained reasonably close to their garrisons, pack animals lacked usefulness and could be troublesome drawbacks, so they were not used. Later, as patrols stayed out longer and moved deeper into the jungle, air resupply by airdrop or airlanding at a jungle airstrip or landing zone became more prevalent. Care had to be taken in these activities that the CTs were not alerted by the presence of hovering helicopters or parachutes stuck in trees. Aircraft also transported some patrols to start positions, reducing the time and effort needed to reach the target area. Aircraft support was relied on more exclusively by deep Special Air Service patrols than by light infantry.

The British also devoted some attention to lightening soldiers' loads while on patrol. To further this end, light but appetizing rations were chosen. Because the soldiers looked forward to breakfast and supper on patrol, the command provided palatable food for these meals.

Other Important Practices

Military operations during the Emergency were also influenced by food control, deception, and route security. Food control was one of the pillars of the Briggs Plan and was vital to the overall counterinsurgent strategy. The aim of this policy was to isolate the guerrillas from their civilian supply sources, forcing them to rely on their own stocks or to move to another area and establish new sources of supply. Once the guerrillas' own stocks ran out (if they did not move), they began to starve and became vulnerable to the "voice" aircraft tempting them to surrender. If they chose to grow their own food in jungle clearings, it made them more visible and less mobile. Once spotted, their jungle crops became targets for aerial spraying, bombing, or investigation by foot patrols. Ultimately, successful food control could starve the guerrillas, who would lack the stamina and the will to stay on the run from healthy, active British patrols.

The British routinely practiced basic food control measures such as licensing sellers and restaurateurs, restricting personal food stocks, and requiring buyers to show ration cards. However, when the decision was taken to mount a major antiguerrilla operation, the British mounted a much more comprehensive food denial program. Food denial took many forms, but its aim was always to squeeze completely dry the daily trickle of supplies to the Communist terrorists in a particular area.

Usually, the British initiated food denial operations by surprise. Thus, when the civilians in an affected district awoke in the morning, they found every gate in the village fence guarded by police and soldiers. In addition, road blocks were emplaced at key points on all roads and vehicular trails. Every person or vehicle (including bicycles and carts) moving through these gates and blocks was searched. "Continual patrolling of the wire fences, day and night... and a dozen other possible methods of smuggling had to be investigated and stopped."[41] Police also arrested all known food suppliers. Meanwhile, emergency restrictions on the cooking of food or on its sale were instituted. For example, every can of food sold during food denial had to be punctured upon sale to ensure its immediate use.

Tedious, unpopular, and wearing, these measures were manpower intensive. In fact, it frequently seemed as if there were not enough men to carry out the tasks. Infantry soldiers, clerks, cooks, and even officers and civilian dignitaries took their turns on the search lines. To facilitate matters, every soldier

Caught red-handed with rice in his bicycle pump

119

was trained in proper search procedures and in the recognition of contraband. These measures were carried out sometimes for weeks and months because it usually took that long for the guerrillas to feel the pinch. During this period, infantry patrols constantly watched the jungle fringe for contacts between civilians and CTs; they also saturated the jungle itself with harassing patrols.

These measures of food denial worked as long as the police and the army applied them rigorously. Eventually, the terrorists either had to leave the area, abandoning the organization it had taken them months and years to build, or, if they stayed, which was more normal, they were forced into dangerous (for them) acts of retaliation or response, such as ambushing food-supply truck columns. But once they emerged from their hideouts for such actions, the British hunted them down remorselessly.

Deception

The terrorists had many sympathizers and supporters in the civilian population, some in places (the Home Guard, telephone service, or the police force) where they could provide valuable information on upcoming British operations. As a result, if information was not guarded carefully, operations often failed. In addition to applying strict rules in regard to operations security, the British also learned to conduct deception to mask coming operations. Often, a great deal of imagination went into these deception activities.

For example, in 1954, the 63d Gurkha Infantry Brigade planned a big operation in the area of Seremban. To disguise it, the brigade contrived a deception plan that they called Operation Whipcord. Whipcord concerned a mythical future operation in a neighboring area, Bahau. To make the plan believable, the brigade ordered maps of the Bahau area and distributed them to units, requested a special rail-loading ramp be built at the Bahau railway station, circulated notices of forthcoming food checks and new regulations regarding food supply, and let slip other related information about the operation. In this way, the Communist terrorists in the real target area were led to relax their guards.[42]

Route Security

For many years, the terrorists were able to interdict the country's road network. In response, the British developed tactical SOPs for route security. Roads were coded according to the threat. "Unrestricted" routes, for example, required no escorts at all; they were considered to be safe. "Black" routes, on the other hand, required an armored-car escort. The SOPs prescribed other detailed procedures in regard to convoy organization, signals, briefings, lookouts, and immediate actions.[43] Violations of the SOPs led to tragedy sometimes as some foolhardy soldier risked the gauntlet.[44] According to Brigadier Henniker, the golden rule for travel was to "demonstrate to any would-be attackers that you would welcome an ambush so as to kill *them*."[45] One conveyed this message by ensuring that every convoy was properly armed, organized, alert, ready, and eager for a fight.

Part II. The Confrontation with Indonesia

Not long after cleaning out the last isolated pocket of Communist guerrillas in Malaya, the British forces in the Far East found themselves facing another limited war, this time in North Borneo. Initially, the threat seemed to be similar to the one experienced in Malaya—that is, a small, lightly armed, indigenous, Communist-inspired insurgency with limited popular support. However, it became clear that the situation in Borneo, (hereafter called the Confrontation) differed significantly from the Emergency in Malaya. Before discussing these differences, however, it is first necessary to describe the background to and initial events of the Confrontation.

Background to the Confrontation

The large island of Borneo in 1962 comprised four political entities (see map 11). Kalimantan, the southern three-fourths of the island, belonged to Indonesia, independent since 1949. In the north, the British administered the two provinces of Sarawak (in the west) and Sabah, also called North Borneo (in the north). The sultanate of Brunei was an independent state ruled by a sultan but possessing a civil bureaucracy and police force staffed to a large degree by Englishmen.

In 1961, Tunku Abdul Rahman, the prime minister of Malaya, proposed the formation of a new federated state to be known as Malaysia. Malaysia, Rahman suggested, should include the Federation of Malaya, the city-state of Singapore, the sultanate of Brunei, and the two colonial provinces of Sarawak and Sabah. Great Britain endorsed the idea, but President Sukarno of Indonesia opposed it, calling it a British neocolonialist project and a threat to Indonesian security. Sukarno had his own dreams about a greater East Asian federation (MAPHILINDO: Malaya, the Philippines, and Indonesia) under his leadership, which also was to include northern Borneo. Sukarno hoped to prevent the formation of Malaysia by using diplomatic, ideological, and if necessary, military means. He openly announced a policy of "confrontation" in January 1963, following the Brunei revolt of December 1962. Thus, the Brunei revolt was actually the opening act in the play that came to be known as the Confrontation.

The Brunei revolt was launched by a small, indigenous, Communist organization—the Northern Borneo National Army (TNKU)—that had ties to the Communist party of Indonesia. Hoping to create a ground swell of popular support for a takeover of the sultanate, the armed rebels achieved a few preliminary successes, notably the capture of the Shell oil fields at Brunei; the towns of Seria, Limbang, and Lawas; and some smaller villages. Possessing few security forces of his own, the sultan requested British military assistance. Within a matter of days, the British deployed by air and sea the 1st Battalion, 2d Gurkha Rifles; the 1st Battalion, Queen's Own Highlanders; the 1st Battalion, Royal Green Jackets; and 42 and 40 Royal Marine Commandos. On 19 December, Major General Walter Walker was appointed director of operations in Borneo. When no popular support for the rebels appeared, Walker's forces quickly reclaimed the facilities and areas taken by the rebels and then, with

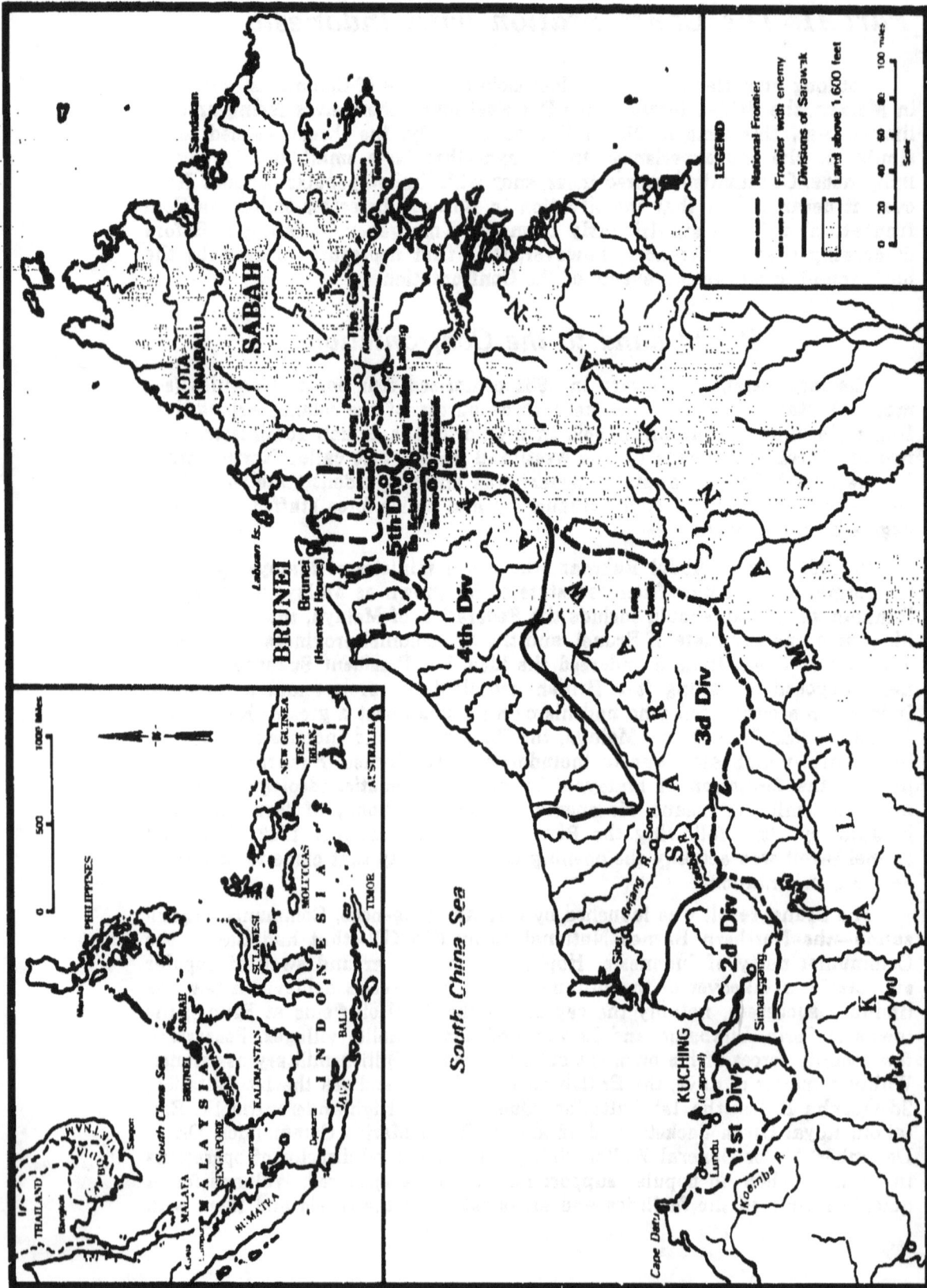

Source: Dickens, SAS: The Jungle Frontier.

Map 11. Northern Borneo

the aid of 2,000 to 3,000 Sarawak irregulars under the direction of a civilian, Tom Harrison, hunted the rebels down in small groups from January to March 1963.

Thus began and ended the first of the three stages of the Confrontation: the defeat of the Brunei revolt and the subsequent mopping-up operations in early 1963. Even though the revolt caught Sukarno by surprise, he seized upon it as evidence of the unpopularity of the concept of Malaysia by the inhabitants of northern Borneo, and he used the revolt as grounds for military support to the rebels and, eventually, for intervention by Indonesian regular forces.

During the second stage of the Confrontation, which took place from April 1963 to April 1964, the Indonesians sponsored periodic raids from Kalimantan into northern Borneo in an attempt to raise guerrilla forces and establish semipermanent camps. In large measure, the early raiders were Indonesian-supported TNKU irregulars, Indonesian-trained guerrillas known as IBTs (Indonesian border terrorists), and some Indonesian "volunteers." When these efforts failed to raise sufficient guerrilla forces, the goal of the periodic raids changed to the creation of destabilization in the border areas. At this time, a sprinkling of regular units from the Indonesian Army began to appear.

The third stage of the Confrontation, characterized by overt Indonesian operations in northern Borneo and Malaya, ran from the spring of 1964 until the end of the war in 1966. During this period, Indonesian regular army units conducted most of the raids into northern Borneo. Indonesian troop strength along the border grew steadily from about 2,500 in mid-1964 to as many as 30,000 in 1965.[46] In response, Walker ultimately controlled four brigades, organized into a varying number of infantry and commando battalions (from ten to thirteen), three to four small Special Air Service squadrons, and supporting air force and naval elements—all of which numbered about 17,000 men at the height of the war.[47]

The terrain and political geography of Borneo gave the Indonesians a tactical advantage from the start. Except for the coastal regions, Malaysian Borneo is a "vast, trackless, rail-less expanse of mountain and jungle."[48] Its primary lines of communication are by sea, river, and air. The hot, humid climate nourishes several different kinds of jungle, among the thickest anywhere in the world, and produces thick morning mists inland and substantial cloud cover.

Surface movement is progressively difficult as one moves inland. Hills rise quickly from the coastal plains and lead to huge mountain ranges covered with thick jungle, which average 5,000 feet in height with some peaks reaching 7,000 to 8,000 feet. In the 1960s, much of the interior was unmapped, and existing maps lacked detail and precision. Borneo's long, twisting 970-mile border with Kalimantan, unmarked most of the way, ran through these uncharted sections. In some areas, no trails or tracks existed; crossing the border in these areas usually meant following a river course.

With such a long, unposted, and unpatrolled border—one immune to air reconnaissance and sparsely settled—hundreds of avenues for incursions into the heart of northern Borneo beckoned to the Indonesians. Moreover, geography permitted the Indonesians to plan and support these incursions in secrecy and safety since the British prohibited cross-border operations and overflights until late in the war.

Comparison of the Emergency and the Confrontation

Was the Confrontation simply a reenactment of the Emergency—the acting out of the same play but on a different stage? Or were the two conflicts unrelated to each other, requiring different methods, means, and concepts? The answer evidently lies somewhere between these views. Striking similarities and important differences existed between the two struggles.

The Confrontation did resemble the Emergency in a number of ways. To begin with, the British used the same basic organizational approach in both situations. As overall director of operations, Major General Walker, in the Confrontation, had broad powers to command not only the army forces committed to him but also the navy and air forces in the theater. He also worked closely with the civil authorities and police in each of the areas, establishing joint headquarters down to brigade and battalion level. This paralleled the earlier British experience in Malaya.

Here, however, a major difference between the two conflicts arises. As high commissioner, General Templer, in Malaya, had directed all civil and military activities. He headed both the civil government and the armed forces. In Borneo, Walker's powers were more circumscribed, since he did not represent the British Crown. Walker, instead of being in charge of the territory, provided only military assistance to the existing governments. The sultanate of Brunei was independent; thus, Walker always had to respond to the sultan as Brunei's head of state. Moreover, because Sarawak and Sabah were administered separately, Walker had to deal with two separate administrations and police forces, each with its own chain of command. As a result, Walker was forced to rely more on cooperation and persuasion than had Templer. When Sarawak and Sabah joined Malaysia in 1965, Walker's situation became even more complex, as he now had to serve a new master in Kuala Lumpur.

The British military forces in Borneo, however, were similar to those used in Malaya. In both locations, light infantry troops formed the core of the committed forces; they were also again organized on an area basis, although the expanse of Borneo required even more decentralization in operations. Troops in Borneo also required the same kinds of skills and tactics as those in Malaya. Furthermore, good, timely intelligence was vitally important in Borneo as it had been in Malaya.

Recognizing that the Confrontation might continue for years, the British adopted a long-term approach to the conflict in Borneo, as they had earlier in Malaya, and resolved to outlast the Indonesians The principal distinction between the two struggles was the limits placed on Walker's power in Borneo owing to the more complicated political structure there. Walker did, however, retain more or less absolute operational control over the army, navy, and air force elements in Borneo.

In regard to the geographical characteristics of the two territories, Borneo's terrain presented the British with more difficult military problems than had

Malaya's. The land was vaster, more sparsely settled, less economically developed, and far more impeding to surface movement. For example, less than 10 percent of the land had been cleared for agriculture or habitation.[49] Furthermore, the high mountains in Borneo had no counterpart in Malaya, and the lack of good military maps posed serious difficulties. Land navigation in Borneo also proved to be much more difficult than in Malaya. These geographical characteristics in Borneo spawned several new operational and tactical requirements: (1) the unpatrolled border produced a greater need for tactical intelligence through ground reconnaissance; (2) the scarcity of troops forced the British to rely on border tribes for information; and (3) because these Bornean border tribes were vulnerable to destabilization, the hearts and minds of the tribesmen had to be influenced positively.

The threat also differed greatly in Borneo from that in Malaya. In Malaya, the Communist terrorists had no real sanctuaries; they had to remain in the country to accomplish their goals. In Borneo, the threat was external, and the Indonesians did have sanctuaries. They could move laterally, attack anywhere across the 970-mile border between Kalimantan and northern Borneo, and retire to safety back in their own territory if the British forces failed to intercept them. Furthermore, the Indonesians operated in larger groups and were better armed, better trained, and healthier than the Communist terrorists in Malaysia. Some of the Indonesians had ever been trained in earlier years at the Jungle Warfare School. Furthermore, some of the raiding parties came from elite airborne battalions. Finally, the Indonesians were more offensive minded. When ambushed or intercepted, they counterattacked. There were no surrendered or captured personnel in Borneo. The enemy in Borneo fought tenaciously and craftily, showing a high level of jungle craft and tactical skill.[50]

Operational Concepts

The British could have made no better choice than Major General Walter Walker as director of operations in North Borneo. Walker's experience and intellect fitted him perfectly for the position. A veteran of the Burma campaign in World War II, first director of the Jungle Warfare School, battalion and brigade commander of Gurkhas in Malaya, Walker had no doubts whatsoever about his fitness to command in Borneo. Described as "the greatest jungle fighter of his time," Walker acknowledged his indebtedness to the examples set by his predecessors, particularly Templer.[51]

From the time his airplane landed in Borneo, Walker knew how he wanted to meet the crisis. His goal was to prevent the escalation of the Brunei revolt and the early Indonesian-sponsored raids into an open war à la Vietnam. To attain this goal, he reasoned that he had to win the opening rounds of the Confrontation and maintain this ascendancy over a potentially long period of time. He concluded, therefore, that the British forces he commanded had to meet each incursion with extreme violence, demonstrating that the smallest violation of the border would result in swift, merciless retaliation against any enemy forces. From this core idea sprang one of the most offensively natured defensive strategies in military history.

Walker established these guiding principles for the prosecution of the war, which he called "ingredients for success":

- Unified operations (i.e., jointmanship).
- Timely and accurate intelligence.
- Speed, mobility, and flexibility of the security forces.
- Security of bases.
- Domination of the jungle.
- Winning hearts and minds.[52]

Unified Operations

Walker believed that the joint organization that he established was indispensable to the successful prosecution of the war. When he arrived in Borneo, he found that the British Army and the Royal Air Force (RAF) occupied widely separated headquarters, and the Royal Navy had no permanent representative ashore. The joint headquarters that he quickly established in one building set the pattern for all lower levels of operation.

Walker's influence on unified operations went well beyond the creation of a joint headquarters. As director of operations, he exercised his full authority to insist that the navy and RAF commanders subordinate their ideas about the proper employment of their forces to his own operational concept. Thus,

Jungle patrol in Brunei

he required the Fleet Air Arm to base its helicopters ashore, and he used the commando ships as ferries to and from Singapore and for local logistical support, not as assault ships. Likewise, he forced the RAF to relax its formal procedures and emphasis on centralized operations. Furthermore, he insisted that it take some risks in the jungle and that they learn new techniques, such as insertion of patrols into the jungle by helicopters using ropes.[53] In this manner, Walker trained the services to approach the war as a team, giving up their parochial viewpoints.

Walker never forgot that military forces in little wars like the Confrontation, as in big wars, existed to support political goals. Though he had no formal authority over the civil and police bureaucracies, he brought them into his joint headquarters and used his influence to obtain their cooperation. He sought also to win their confidence through tactical successes and effective civic actions in the border regions. The British did experience some problems with the police in Sarawak and Sabah.[54] Nevertheless, a spirit of cooperation generally prevailed between the civil, police, and military arms.

The Domination of the Jungle

Tactically, the most important of Walker's six principles was his order to the infantry to dominate the jungle. This principle grew out of Walker's attitude of offensiveness. The principle was tied inextricably to the idea that every Indonesian incursion would be met with violence. Dominating the jungle meant making the jungle one's home for weeks on end. It meant, in some respects, living like a guerrilla and using one's primitive instincts and senses—becoming a jungle creature hunting for its prey. More than anything else, domination of the jungle required a frame of mind that accepted the rigors and dangers of life in the jungle and determined, at the same time, that the jungle could be used to one's advantage. The jungle was to belong to the British, not to the enemy—that was the theme.

Operationally, Walker demanded that the British maintain a continuous, shifting presence along the entire border through constant patrolling. In this way, the British aimed—primarily through the means of ambushes—to create a strong sense of insecurity in the minds of the enemy, a sense that by crossing the border they were putting themselves in great peril. Perhaps the Indonesians might not be intercepted on the way in, but once their presence was detected, they knew that the British infantry would doggedly pursue them every step of the way thereafter. The enemy would be given no respite, no chance to relax. In this manner, through sacrificial effort, the British would maintain ceaseless pressure and relentless pursuit of the foe.

The history of the Confrontation is replete with examples of how Indonesian raiding parties, over weeks, were hunted down to the last man. For example, in December 1963, a 128-man enemy force raided Kalabakan in Sabah. Repulsed by the local forces, the Indonesians lingered before returning to Kalimantan. In a flash, the 1st Battalion, 10th Gurkha Rifles, had cut them off and begun pursuit. By the first of March, the Gurkhas killed or captured 96 of the 128 enemy soldiers.[55] In late 1963, in Sarawak, after another failed raid at Song, the Gurkhas harried the Indonesians for a month as they tried to withdraw.[56] Few enemy hit-and-run sorties recovered to their own areas without suffering significant casualties. Retaliation by the British could not be avoided.

Dominating the jungle boiled down to the question of who would be the most aggressive, the most dangerous, the most ruthless—the British or the Indonesians. It was a contest for mastery of the jungle. In the end, the British showed more cunning, guile, craft, discipline, and sacrifice than did the Indonesians, so they beat them in the deadly game of jungle ambush and retaliation.

Speed, Mobility, and Flexibility

Walker's operational concept dictated that his forces respond immediately to every hostile enemy action. Forward deployment of his forces and decentralization along the border established the framework for an immediate, flexible reaction. Unfortunately, Walker never had enough infantry forces to do anything but maintain the thinnest of screens.

Brigade frontages were enormous, varying from 81 miles in the most threatened area, to 442 miles in another. Within a brigade, individual battalions assumed responsibility for vast areas. In 1964, for example, the 1st Battalion, Royal Leicestershire Regiment, covered a front longer than that of the British Army of the Rhine and an area the size of Wales. Moreover, platoons and detachments were as much as 100 miles from any permanent base. Table 5 below shows the distance in miles between individual elements and the battalion headquarters.[57]

Effective domination of the jungle, in the view of such extreme decentralization, depended, among other things, on the capability of forces to react rapidly to the discovery of the enemy. Clearly, dominating the jungle where the enemy was not had little value. The light infantry had to get to where the enemy was before he could retreat and escape. The solution to this problem was to obtain early warning of the enemy and to achieve speed, mobility, and flexibility in his pursuit. The latter principle was fulfilled, above all, by the use of helicopters.

The use of helicopters permitted Walker to implement his plan of forward deployment. The numerous permanent jungle bases constructed within a few kilometers of the border at various widely separated points along its length

Table 5. The Disposition of the 1st Battalion, Leicestershire Regiment, 1964

Unit	Location	Distance from HQ
A Company	Tawau (Sabah)	250
B Company	Bangar	20
C Company	Lawas	85
5th Platoon	Ba Kelalan	80
9th Platoon	Long Pasia	70
10th Platoon	Pensiangan	100
11th Platoon	Sepulot	100

128

relied exclusively on the air line of communication. All supplies except water arrived by helicopters or airdrops. The cargo was sometimes quite unusual. An Australian company commander having trouble with rats requested "cats, pussy, 12" and got them.[58]

The most important contribution of the helicopters was the tactical mobility that they provided. Initial skirmishes with the enemy had shown that sending a force on foot to a hot spot simply took too long: by the time the British infantry arrived, the enemy had disappeared. Through the use of helicopters, however, a relief or ambush force could be in place in a matter of minutes. Moreover, these forces could be emplaced in several places, cutting off the enemy regardless of his direction of flight.

The basic procedure for immediate reaction was called the Step-Up Drill. To effect this drill, each battalion maintained an on-call force in combat readiness adjacent to a pick-up zone. Whenever an Indonesian force was spotted, this on-call force, alerted by radio, mounted its transports and promptly flew in to the nearest landing zone. The SAS conducted this drill for village headmen in remote areas to demonstrate to them that even though no large troop unit had bases nearby, a substantial force could be flown in to protect their villages in mere minutes.[59]

New techniques enhanced these rapid moves. For example, in March 1965, a 150-man Indonesian company attacked the platoon manning the base of B Company, 2d Battalion, The Parachute Regiment. Though supported by engineers and rocket launchers, the attack failed. The British followed up on the attack almost immediately. Three platoons with five days' rations roped down from helicopters into the jungle behind the enemy to set up ambushes, and A Company was flown in to the B Company base to pursue directly. In this case, the surviving Indonesians escaped without further losses; still, the methods of immediate reaction were sound.[60]

In another case, the enemy was not so lucky. When a fifty-man element penetrated the frontier in August 1966, an entire battalion was deployed to round them up. Operating in platoon packets over about 200 miles of territory, the battalion annihilated the enemy force to the last man in a month's time.

Naturally aircraft could be used to transport troops to anywhere within their range. Consequently, a battalion did not have to rely on its own resources to meet a threat. A neighboring battalion might provide the reaction force. If a reaction force stayed out for an extended period of time, it was resupplied by air.

The British took the burden off the helicopter force by building jungle airstrips for light aircraft, such as the Beaver, which were used to transport troops and cargo. The British also cleared several hundred loading zones along the frontier, which allowed them easy entry to hot spots. (Local natives often helped clear these loading zones under direction of the light infantry or the SAS.) Any of these loading zones could also be used to pick up patrols and evacuate casualties.

The supporting helicopter squadrons came from the RAF and the Fleet Air Arm. Some of the unit commanders hesitated initially about deploying helicopters so far forward, but as the methods prescribed by Walker and his subordinates proved effective, commanders became enthusiastic supporters. Nonetheless, the weather often restricted operations, and air navigation was difficult. Furthermore, it was easy for pilots to get lost, unless they had experience in the area of operations. Through necessity, pilots navigated primarily by terrain association and dead reckoning (that is, by using timed flights at fixed speeds along fixed headings). Few in number, the helicopters were controlled centrally but deployed widely.

The helicopter force was of utmost importance to operations. Walker's estimate of their value is evident from his statement that an infantry battalion with ten helicopters was worth more than a brigade on foot.[61] Even more telling was his refusal to accept more infantry units unless he also received an increase in helicopters. The use of helicopters gave the British an advantage in tactical mobility while neutralizing, to a certain degree, the Indonesians' freedom of maneuver once they crossed the frontier. In summary, through their adroit use of helicopters, the British defeated the enemy even though they were outnumbered by him and on the defensive.

Security of Bases

Whether the Brunei revolt would develop into a larger, wide-scale insurrection was unclear at the beginning of the insurrection. One estimate concluded that there were 60,000 potential guerrillas in North Borneo.[62] Because of this threat and Walker's principle that all military and police facilities had to be able to protect themselves wherever they were, potentially every soldier and policeman might become involved in the defense of their garrisons. Walker's concern proved well founded when during the first year of the war, the Indonesian-supported TNKU guerrillas attacked a number of installations deep inside the frontier. The seriousness of this threat diminished, however, as the British increased the size and training of the police force and as the guerrilla force faded away. From 1964 to 1966, the forward jungle bases and border villages were most threatened by Indonesian regular forces, not the interior garrisons.

To meet this threat at its source, the border, Major General Walker directed the construction of jungle bases well forward. These bases, it was hoped, would deny the enemy access to the interior of North Borneo and provide the British with a variety of advantages. They functioned primarily as widely separated secure havens for the men conducting constant patrolling in the frontier zone and afforded a place for returning patrols to rest, relax, eat hot food, and take hot showers. In addition, the bases protected nearby villages by virtue of their proximity. They also served as a focal point for the collection of intelligence from the local natives. Furthermore, units based in these jungle forts carried out civic-action programs in nearby villages. The bases were never meant to serve as static defense forts; that kind of strategy was doomed to failure.

The organization of jungle bases varied somewhat, but it normally included an infantry company, a mortar detachment, a landing zone, an artillery section of one or two guns, and living space for extra forces if needed. Occasionally, the base included a helicopter detachment as well. The base was manned by one platoon on a rotating basis; the other three platoons stayed in the jungle hunting the enemy. Walker and his subordinate commanders refused to let the need to defend the base compromise the order to dominate the jungle.

The British made no attempts to hide the forts. Constructed on high ground and fortified with trenches, sandbag bunkers, wire, punji stakes, claymore mines, and overhead cover, the forts were formidable positions. The primary defensive weapon system was the tripod-mounted medium machine gun, supported by mortars and, when on hand, a 105-mm gun in a direct-fire role. The vegetation around the perimeter was cleared to improve fields of fire, and some forts put up lights for illumination at night. Sentry dogs enhanced early warning of enemy approach. The Indonesians tried several times to destroy some of these bases but never succeeded.

Timely and Accurate Intelligence

Early warning was critical to the success of British operations as they have been described above. Indeed, without early warning of Indonesian intrusions, the entire defensive scheme designed by Walker could only fail. Unfortunately, the requisite intelligence infrastructure to support early warning did not exist. The several police forces that could have provided intelligence were small, and there was no Special Branch. Furthermore, in the frontier areas where the greatest threat existed, police posts and villages were separated by tens of miles of daunting terrain. Consequently, information regarding the Indonesians came from two primary sources: the border tribes and the armed forces themselves.

The border tribes possessed an immense potential for intelligence collection. Adept in the jungle, they easily concealed themselves from the British and the Indonesians. Their hunting forays often brought them into contact with enemy patrols. Moreover, many had relatives or trading partners in Kalimantan, so they had valid reasons for crossing the border. However, obtaining information from the border tribes depended on the ability of the army to protect them from Indonesian raids. Isolated in their village longhouses, the highland aborigines traditionally were favorably disposed toward the British because of the peaceful and beneficial colonial heritage. However, experience proved that they would not help the British unless they were sure of protection. In several instances, the local natives were aware of cross-border movements by the Indonesians, but they did not notify the British forces or the police because they feared retaliation.

As a result, the British devoted a great deal of effort to convincing the border tribes that they could protect them. To secure native confidence, British security forces maintained a frequent and visible presence. Special Air Service patrols, in particular, lived in many of the isolated villages, where they endeavored, through staged Step-up Drills and their own fearless patrolling, to win the trust of the people. If a village was known to be threatened or victimized, the British immediately sent a formation to its aid. In the process, they paid proper respect to the village headmen by listening to their concerns, responding to their requests, and visiting their longhouses frequently. Villagers and headmen received advice and support on their own self-defense as well.

The British were also exceedingly careful in their own operations not to endanger civilians. The procedures established for calling for close air or artillery support, for example, had tight restrictions to prevent civilian casualties. In addition, the British avoided pitched battles for the control of villages; whenever possible, they confronted the enemy before he reached target areas. The security forces were amazingly successful in this regard. From 1965 to early 1966, the British defeated more than 200 separate enemy operations. Only four of these Indonesian raids penetrated to within mortar range of their objectives.[63]

The Special Air Service, using the skill and daring of RAF and navy crews

Reprinted by permission of Arms and Armour Press

The campaign to win the hearts and minds of the people also contributed to the willingness of the people to come forward with information. Gradually these measures persuaded the border people that their security would be improved through their participation with the British as sources of information.

Some of the members of the frontier tribes went beyond this passive participation. These men were the Border Scouts, an irregular force numbering about 1,500. Initially conceived as a paramilitary self-defense force, an idea that did not work, the scouts were turned instead to reconnaissance and intelligence—roles for which they were well suited. The scouts received fundamental training and guidance from the infantry and SAS units with which they were associated. Moving freely between villages and across the border, the scouts collected information on the locations and movements of Indonesian units and their armament, size, and disposition. They reported this invaluable information back to the armed forces, which were then able to plan and act accordingly. The raising and use of the Border Scouts helped to create that fine intelligence mesh so necessary to rapid reaction by the British.

The British also relied on their own forces to provide intelligence—their principal instrument being the long-range reconnaissance patrol (LRRP). During the conflict, there existed a mild debate within the army concerning who should conduct long-range patrols, the light infantry battalions or the Special Air Service. Operations demonstrated that both could be successful but that only the most experienced and able troops should be employed in this task.

All of the British, Gurkha, and Commonwealth infantry battalions sent to Borneo formed their own LRRPs. These patrols varied in size but generally operated in small groups. They used local guides (Border Scouts) and stayed in the jungle for weeks, resupplying by air and operating as much as 120 miles from base. The primary mission of these LRRPs was to collect information on terrain, local conditions, routes and trails, and enemy locations, movements, and activities. In certain situations, notably self-defense, they were permitted to engage in offensive action. Generally, however, they scrupulously avoided detection, denying themselves the pleasure of hitting a fat target in order to keep the information flowing into headquarters. The patrols operated in the most remote areas. Even if they did not make contact with the enemy, the information that they collected on the terrain had great value.

Conducting such long-range reconnaissance undoubtedly was the most demanding task performed by the light infantry in Borneo. Deep in largely uncharted territory, dependent on an air lifeline, miles from the nearest help, the patrol members of the LRRPs had to be the most stable and capable of men. Their task required the highest standard of jungle craft and strong nerves.

Lieutenant Colonel H. J. Sweeney, commander of the 1st Battalion, Royal Green Jackets (an outstanding battalion), insisted that his LRRP be composed of the best soldiers in his battalion. As shown in table 6, he prescribed the necessary characteristics of each individual and the special training that was needed.[64]

Table 6. Necessary Characteristics and Training, 1st Battalion, Royal Green Jackets

1. First-class shot on all weapons	1. Jungle marching, carrying radios and four days' rations
2. Tough and resilient	2. Watermanship; operation of outboard motors
3. Above-average intelligence	3. Living in the jungle for long periods of time
4. Well-balanced infantry skills	4. Use of radios and Morse code
	5. Colloquial Malay
	6. First aid

Particularly noteworthy are the last two areas included in their suggested training: language skills and first aid. The LRRPs often visited remote villages to obtain information, demonstrate British presence, and establish friendly relations. It was essential that all, or some of the group, be able to speak the language. The group members also had to be able to minister to their own and their comrades' wounds and illnesses and the maladies of the aborigines whom they visited. Sharing British medical aid with the tribesmen built strong bonds of friendship and trust.

While the many infantry battalions in Borneo sent out their own LRRPs and directed significant intelligence collection activities through Border Scouts, village visits, consultation with police, and local patrolling, most of the deep patrolling was performed by the SAS. The 22d SAS Regiment entered the Confrontation soon after the Brunei revolt. Initially, a danger existed that the SAS might be used in a reserve strike role. However, the SAS commander, Lieutenant Colonel John Woodhouse, convinced Major General Walker that the SAS should be Walker's "eyes and ears." He insisted that Walker use them "to establish a forward deployed intelligence/communication net right in the jungle with the natives near the border."[55] While the SAS is a very specialized form of light infantry—high above the norm in terms of training, skills, and capabilities—it embodies, in many respects, the high standards to which all light infantry units should aspire. For this reason, its operations in Borneo should be examined.

The Special Air Service

After honorable service in World War II, the SAS had been resurrected for the Malayan Emergency, during which it pioneered the roles and tasks it was later to assume in Borneo.[56] In Borneo, it was employed on the outer edge of the British defenses to act as a trip wire in providing early warning of Indonesian incursions. Its role was entirely one of watching and reporting. Only in rare situations did the SAS engage in offensive actions during the first months of the Confrontation.

With Walker's blessings, Woodhouse deployed his SAS squadrons in four-man teams across the entire frontage. Each team moved into a native village for a fixed four-month tour. Sharing the dangers and the communal life of

134

these frontier aborigines, the SAS teams ate, worked, and slept with them, winning their friendship and respect. Speaking the local language, these SAS soldiers were able to share their technical skills—especially medical help—with their hosts. By winning the hearts and minds of the aborigines, a fundamental ingredient in the SAS repertoire, they were able to rely on the villagers for information and cooperation in the fight against the Indonesians.

Using the local village as their base and moving from one village to the next, the SAS painstakingly reconnoitered the entire border. Thus, one of their earliest contributions to infantry operations was the compilation of terrain descriptions and sketch maps for these unmapped areas. The teams remained in the same general area during these exacting four-month tours in order to become thoroughly familiar with it. SAS squadron commanders each went forward on foot to personally reconnoiter the territory in which their teams operated. One officer walked almost the whole 970-mile border.[67] Later in the war, as the infantry's forward presence increased in size and skill, the SAS teams moved across the border to observe the enemy and to perform terrain reconnaissance. These tension-filled cross-border reconnaissance patrols then became the norm for the SAS.

Using four-man teams in this manner, with assistance from local Border Scouts, the SAS left few gaps through which the enemy might slip unnoticed. The SAS maintained close liaison with the infantry, often by assigning their wounded and ill as liaison officers. When they detected enemy movement, they reported immediately to the infantry, who then reacted with a Step-Up Drill to intercept the Indonesians. So effective was the SAS in providing early warning that Major General Walker remarked that "I regard 70 troopers of the SAS (one squadron) as being as valuable to me as 700 infantry in the role of hearts and minds, border surveillance, early warning, stay behind, and eyes and ears with a sting."[68]

The SAS succeeded because they possessed extraordinarily high levels of combat skills and field craft, levels rarely reached by even the best light infantrymen. The three most important skills have already been cited: radio communications, first aid (the equal of the average medic), and language. Extraordinarily fit, these specialists practiced marksmanship and quick fire with a dozen different kinds of weapons. They also trained in parachuting (into water and trees), abseiling, demolitions, booby traps, survival, and the use of all kinds of vehicles and water craft.

The SAS approached the standards of the aborigines in jungle craft and tracking. Just plodding along through the jungle was not enough. Endurance was essential for SAS patrols, along with meticulous attention to detail. Isolated and exposed, under constant nervous stress from the danger of detection, the SAS teams had to be keen observers, anticipating, making minute decisions, choosing the best routes, eking out their rations, and measuring options in the event of an emergency. As much as possible, they left no indications of their own passing or presence. Moving silently through the bush, the SAS strained to recognize the signs of the enemy: urine stains, bruised moss, machete marks, cigarette papers, and footprints under leaves or along river banks.

As navigators, the SAS was unequaled. Its methods depended on close scrutiny of aerial photographs, memorization of recognizable terrain, detailed note taking when on the move, and strict measurement of distances and bearings. The SAS was able to call helicopters to within several hundred meters of its positions for resupply or evacuation despite having been on the march for a week or more in unknown territory.[69]

The SAS adhered rigorously to a number of field-proven SOPs. For example, the contents of rucksacks were regulated according to men's specialities (radio operators, medics, etc.) and were weighed to ensure they did not exceed fifty pounds. Personal loads were also spelled out for each operation, as shown below in the preparations made by a four-man patrol led by Captain England:

> Secrecy, security, self-sufficiency, and deniability were England's watchword in making his plan, which began with what each man was to carry and where, in minute detail. In his hand he would carry his self-loading rifle with full magazine. On his person would be his escape compass, 100 Malay dollars sewn into his clothing for soliciting help in emergency, field dressings, morphine, plasters, torch, notebook and pencil, map (never to be marked with his true position, but a fictitious track entirely in Sabah to imply a genuine navigational error), loo-paper, matches, knife, watch, and wrist compass for those lucky enough to own one. On his belt would be his compass, 'parang' (machete), two full magazines, water-bottle, mug, sterilizing tablets, two days' rations in his mess tin, spoon, cooking stove with hexamine fuel tablets, more matches, paludrine (malaria pills taken daily), wire saw, insect repellent, rifle cleaning kit, and a hand grenade.
>
> The bergen's contents varied from man to man. Hoe was the signaller, carrying the radio with its spare battery, aerial and codes, and the Sarbe (radio beacon), which had at last been issued. These were heavy items, so most of his food was shared around the others, leaving him with his spare shirt, trousers, socks, boots, poncho, sleeping-bag of parachute silk, nylon cord for contingencies, and book for beguiling the hours when not on observation duty, though not during the eleven-hour nights when the escape would have been most welcome; a candle on a sharpened stake conveniently positioned at the hammock-side was a luxury of the past, for no lights or stake-sharpening would be permitted now. Condie's extra load was the medical pack, containing surgical scissors, forceps, thermometer, syringe and needles, scalpel blades, suture needles and thread, extra morphine syrettes, sterile water, assorted plasters and bandages, and a comprehensive pharmacy. England and Manbahadur took the binoculars, camera, and two large water-bags. The latter were carried empty; on passing a stream all would replenish their personal bottles and drink their fill—and more, for one can never have too much in the tropics, while too little causes heatstroke which can kill as readily as hypothermia. A night-stop near a stream would not be safe and it was then that plenty of water was needed for brewing, soaking dehydrated foods, cooking and washing-up. The supply was carried up in one load and the waterpoint never used again. Rations were keenly debated and whittled down, for it would surely be acceptable to lose weight for a maximum of twelve days rather than carry an incapacitating and tiring load.[70]

The SAS also maintained SOPs for ambushes, immediate actions, and escapes and evasions. However, it showed proper flexibility in its willingness to change these SOPs when circumstances or experience showed that change was needed. In the words of one SAS veteran, the unit was a great "finger-poking regiment," meaning that each patrol or operation, no matter how successful, was subjected to intense examination by its participants and other SAS members in order to determine in detail how it might have been done better.[71]

136

Sergeant "Gipsy" Smith's hydroelectric generator at Talinbakus, Sabah

The SAS in Borneo was characterized by the highest standards of self-discipline and field craft, resistance to mental stress, relentless pursuit of excellence in its operations, and dogged perseverance in going one step further than required. The SAS exhibited great confidence in itself.

Winning Hearts and Minds

During the Vietnam War, the concept of winning the hearts and minds of the inhabitants of a country threatened by an insurgency became very much a part of the American vocabulary. However, it was the British who originated this concept in Malaya and then implemented it to near perfection in Borneo. That winning the hearts and minds of the people in Borneo implied maintaining their security is clear. The British accomplished this task through constant patrolling, deterring enemy attacks, immediate reaction, direct help, and assistance in village self-defense.

Although the British soldiers were technologically superior to the Bornean natives and far more sophisticated and educated, they took care to treat the people with respect in order to win their cooperation. They did not adopt the attitude of a conquering or occupying army. They approached the people as equals and meted out fair, kind treatment in all matters of mutual interest. In particular, the village headmen were given the honors befitting their positions. Commanders did not dictate to the people; they consulted with them and explained their operations and policies. Moreover, soldiers observed local customs and adhered to rigorous guidelines for behavior when in contact with the natives.

137

Officers and soldiers also showed the natives respect by learning the local language (a lesson often lost on the U.S. Army even today). The SAS went one step further by actually living with the villagers, eating their food, entering into their celebrations, and making real friendships. By using the villagers' language and by sharing their life-style, the British made a favorable impression on the natives.

The most important service provided by the army to the Bornean tribesmen was medical aid. No other act of aid given to the tribesmen by the British compared to that of mending a broken limb or administering the medicine that cured a dangerous disease or corrected a long-standing illness. All four-man SAS teams included a man with extensive medical training for just this purpose. Infantry medics performed similar services on every visit to a village or jungle longhouse.

Finally, the infantry participated in many civic-action programs, such as skill training, local construction, improvements in local agriculture and water supply, or arranging the delivery of needed materials. The motive behind such programs was to foster self-sufficiency in the natives. Soldiers guided, advised, and assisted the villagers; they did not do the work for them.

"Winning the hearts and minds": a medic at work

138

The result of these efforts was apparent. The implementation of the principle of winning hearts and minds in Borneo was influential in obtaining the concrete assistance provided by the people to the army in the forms of information and early warning, in the natives' willing help in building landing zones and clearing trails, in the performance of the villagers as guides, and in the formation of the tribesmen into the essential Border Scouts. The British were fortunate in that their colonial heritage and Malayan experience enabled them to be sensitive to the well-being and attitudes of the people and to capitalize on their good will.

Claret Operations

During the first twenty months of the Confrontation, the political and military situation in Borneo led the British government to dictate severe restrictions on the use of force against the enemy, primarily to keep the conflict from escalating and to demonstrate clearly that the Indonesians were the aggressor. However, by September 1964, Walker's forces had successfully stabilized the situation. The security forces on the frontier had proved to be more than a match for the Indonesians, while the British air and sea forces discouraged enemy strikes in those quarters. At this juncture, Walker was able to secure permission to mount some discreet cross-border operations called Claret operations.

The Claret operations had a specific goal—the creation of a defensive attitude in the minds of the enemy. Through a series of shallow, sharp, and violent raids into Kalimantan, Walker hoped to put the Indonesians off-balance and to intensify their feelings of uncertainty. Previously, the enemy had had every reason to feel safe and secure as long as it stayed on its side of the border. Claret operations were designed to erase that sense of security and replace it with a feeling of uneasiness. Walker believed that well-placed pinpricks could make the whole enemy body tremble. The operations were to receive no publicity because of their sensitivity. Claret operations were regulated by strict guidelines known as the "Golden Rules":

1. Major General Walker was to approve all "Claret" raids personally.

2. Only trained and tested troops could be used. No troops in their first tour of duty were to participate.

3. Depth of penetration was to be limited.

4. All attacks were to have the specific aim of deterring enemy offensive action and must never be in retaliation or simply to cause casualties.

5. No operation requiring close air support could be undertaken. Helicopters were not permitted to enter enemy airspace except in dire emergency (by approval of the director of operations).

6. Each operation must be planned with a sand table and rehearsed for at least two weeks.

7. Absolute operations security was necessary. Full cover plans must be made and all nonessential items traceable to the British forces must be left behind (including, for example, dog tags).

8. On no account must any soldier be captured either alive or dead by the enemy.[72]

The first Claret raids were shallow affairs carried out by superb Gurkha troops and the SAS. The SAS participated in the raids since it knew better where the Indonesians were and could guide the Gurkha units to their objectives. The SAS, permitted to run its own four-man raids, was gleeful about the chance to do something other than "watching and counting." The SAS nicknamed itself "The Tiptoe Boys" because of its ability to strike the enemy hard and slip away, leaving only empty space to receive the enemy counterattack.[73]

Infantry raids still assumed larger proportions (up to company size) than SAS raids, but they remained small enough to avoid precipitating a violent Indonesian response. As the risks were great, so were the precautions: "no rifleman was allowed to eat, smoke, or unscrew his water bottle without his platoon commander's permission. At night, sentries checked any man who snored or talked in his sleep. Whenever the company was on the move, a reconnaissance section led the way, their packs carried by the men behind."[74]

Appearing ghost like out of the jungle, these parties of light infantrymen usually achieved complete surprise. After conducting a trail or river ambush or an early-morning attack on an Indonesian border post-forward base, the Gurkhas immediately returned to the friendly side of the border. Pursuing Indonesians had to take care to avoid being caught in an ambush.

As time passed, more and more of the experienced infantry battalions were given permission to participate in Claret operations. Walker also increased the depth of penetration from 5,000 yards to 20,000 yards. Ultimately, the operations accomplished their goal: the frequency of enemy offensive actions in Borneo fell off as the Indonesians became preoccupied with protecting themselves. Throughout the war, the British never acknowledged their raids into Kalimantan territory. The Indonesians, on their part, were embarrassed too much by the raids to make a political issue of them.

Tactical Issues

The tactics, combat support, and individual skills required by the British in Borneo resembled those practiced by their light infantry in Malaya. Nevertheless, because three years had elapsed since the struggle in Malaya, the troops initially deployed to Borneo required extensive acclimatization to the jungle climate and retraining in jungle warfare. As in Malaya, infantry units received their training at the Jungle Warfare School, where they rapidly reacquired the necessary skills.

Significant differences did exist, however, between the tactical style of operations used by the British in Borneo and that in Malaya. Owing to the larger size of the threat in Borneo (the Indonesians rarely moved in groups smaller than a platoon), the British infantry generally operated more at the platoon level than at the squad level. In addition, because the enemy fought more tenaciously in Borneo, the British devoted more attention to conservation of ammunition. Isolated patrols could not afford to run out.

Light infantry attacks in Borneo most often took the form of ambushes. These ambushes, as in Malaya, lasted for long periods of time. One infantry unit, for example, maintained an ambush in waist-to-shoulder-deep water for three days, rotating the men on the ambush site every ninety minutes.[75] In Borneo, the infantry conducted more river ambushes than they had in Malaya. The sparse settlement in Borneo also permitted the British to set up remote ambushes using claymores and other mines—ambushes that were self-detonated by the victims. Occasionally, these mines were triggered by animals. However, even if the Indonesians did not fall victim to these remote ambushes, their nondetonation informed the British that no enemy patrol had passed that way.

Close-air and artillery support were used more widely in Borneo than in Malaya, mainly because the Indonesians presented better targets. To support local patrolling, 105-mm howitzers were deployed into jungle bases singly and in pairs. Using these widely separated pieces was not easy, however, because of the difficulty in controlling indirect fires. Infantry NCOs had to be proficient in calling for fires, since they seldom had forward observers along. Similarly, each gun section had to have the capability to compute firing data. In addition, a special fire-control net was established. In view of the extraordinary requirements of the situation, the artillery command in Borneo published special, area-specific SOPs.[76] The RAF conducted no bombing operations in Borneo.

Logistics

The greater difficulty of the terrain, the lack of a decent road net, the wider decentralization of forces, and the improvements in helicopter technology and techniques influenced the British to provide 90 percent of their logistic effort in Borneo by air. The British also employed watercraft in resupplying units. The Hovercraft, in particular, was put to good use, carrying both troops and cargo via inland waterways. Watercraft, unlike helicopters, could also operate at night.[77]

Owing to its precarious situation, the SAS supplied itself by placing caches here and there for emergency use. It also supplemented its light rations with jungle foods such as fruits, bamboo shoots, animals, and other local fare. The jungle could have sustained the SAS completely, but such an approach would have consumed too much of its time. The standing requirement for SAS teams on patrol was to be able to vanish for two weeks without having to resurface for resupply.[78]

Weapons and Equipment

The mild controversy over weapons and equipment that was generated during the Emergency grew in intensity during the Confrontation. Walker considered that none of the available standard infantry weapons was satisfactory.[79] The issue rifle (SLR) was too long and heavy for use in the jungle. Troops much preferred the AR-15, the export version of the U.S. M-16. Its lighter ammunition (5.56-mm) and high velocity seemed much better suited to their situation than the slower, 7.62-mm NATO round. Shotguns, once again, demonstrated their great utility for close fighting in heavy vegetation.[80]

The British reevaluated their use of other weapons. The aging, but highly regarded, Bren gun was in the process of being withdrawn from the inventory during the 1960s, but its replacement, a belt-fed medium machine gun was

141

A joint planning session of SAS and cross-border scouts

deemed too heavy and too susceptible to malfunction from dirt and water. In the 1950s, the ATOM manual declared that the 2-inch mortar had no utility in the jungle; nevertheless, in Borneo, the 81-mm mortar was thought to be too heavy for patrolling, and units preferred the 2-inch mortars. Its reduced range posed no problem in the close quarters of jungle warfare.[81]

Walker complained to his suppliers that many other items of equipment weighed too much for light infantry work. He identified tactical radios, air-ground radios, jungle clothing, and rations as items requiring lightening.

Significance of the British Victory in Borneo

The significance of the British victory in Borneo has been largely over-looked by analysts, doctrine writers, and military planners in the United States. The widening conflagration in Vietnam eclipsed it on the world stage at that time. Yet the British deserve praise for their accomplishments. Outnumbered by the enemy and on the defensive, suffering from significant tactical dis-advantages, severely restricted by the international anticolonial climate of the 1960s, and restrained by limited national political objectives, the British Army fought a three-year and four-month-long campaign in a hostile environment, winning by virtue of their superior organization, leadership, technology, and light infantry tactics. As a case study in protracted low-intensity conflict, the Confrontation has few equals. With justice, Denis Healy, English secretary of state for defense during the war, said that "in the history books it will be recorded as one of the most efficient uses of military force in the history of the world."[82]

Conclusions

The British operations in Malaya and Borneo had their own unique sets of conditions and parameters that strongly influenced the flow of events as well as the methods and techniques employed by British military forces there. One must be careful about indiscriminate identification of lessons from the operations. Nonetheless, certain tactical and strategic principles were employed in the two conflicts that characterize light infantry operations in general.

At the tactical level, the most important principle to be inferred from the British experience is that light infantrymen must be masters of their environment. In Borneo and Malaya, infantrymen had to be willing to live in the jungle under the most primitive conditions. They had to endure grueling terrain, exhaustion, heat and humidity, jungle pests, and severe mental and physical strain just to be able to get at the enemy. Moreover, they had to be more adept and capable in the jungle than their enemies. Had the Communist terrorists or the Indonesians been able to dominate the jungle rather than the British, the outcomes of the conflicts would have been far different. Indeed, Walker's maxim to his forces to "dominate the jungle" should be taken one step further: light infantry must dominate the environment whether it be jungle, mountains, or arctic wastes.

To achieve this mastery over the environment requires, above all else, a singleness of purpose, an attitude of self-reliance, and an unflinching mental discipline and self-denial. Furthermore, such domination of the environment depends on the adoption of an offensive, aggressive policy. For this reason, the light infantry in Borneo and Malaya constantly endeavored to wrest the initiative from their enemies. Walker's Claret operations are a perfect example of an offensive orientation within the context of an overall defensive strategy.

In both conflicts, the development of timely, accurate intelligence led to the success of tactical operations. This intelligence originated in large measure from the local inhabitants and the police organizations. Good intelligence appears to be an indispensable cornerstone for light infantry operations in counterinsurgency and low-intensity conflicts. The link between tactical success and the free flow of public information is undeniable.

The British record in these operations also demonstrates that regular conventional infantry (including conscripted troops) can be employed as light infantry provided that the leadership and the rank and file are given proper, extensive training. This process takes time. Commanders in Borneo testified that even after completing the Jungle Warfare School, good infantry battalions still required four months' experience in actual operations before they began to be effective.[53] Infantrymen had to be acclimatized before their tactical proficiency began to improve.

The small-unit actions of the Emergency and Confrontation placed a premium on the quality of low-level infantry leaders. The burdens of responsibility and decision making borne by the NCOs, lieutenants, and captains in these campaigns far exceeded that experienced by the same ranks in conventional large-unit operations. Because the demands placed on light infantry leaders in such operations are great, one must conclude that the abilities of such men should be commensurately high.

While leadership was crucial to British success, superior technology also played a part in their victories. But technology was never allowed to dictate the terms of the battleground. Rather, the primary theme of British combat was always to close with the enemy on the ground and defeat him with infantry and close infantry weapons. Combat support systems served to support this primary aim, but infantrymen were the decisive weapon system.

Even though light infantrymen carried the brunt of combat in Borneo, there was a limit to the physical and mental strain that could be placed on them. For this reason, SAS squadrons were restricted to four-month tours followed by rotation. Their commander, Lieutenant Colonel Woodhouse, adamantly refused to consider extensions. Beyond this time, like the earlier Chindits, the SAS teams rapidly lost their effectiveness as their physical and mental edges were worn dull. Similar policies were followed in regard to the periodic relief of infantry battalions deployed along the frontier. The situation was handled differently in Malaya. There, the return to garrison by soldiers for four- to five-days' rest before they began their next operation restored their physical strength and renewed their mental sharpness.

The actions in Malaya and Borneo paralleled each other in important respects. For example, the British adhered to a 3x3 squad organization in both conflicts. This type of organization appeared to have high value in independent squad-level actions, probably because of the flexibility that it provided. In addition, security forces in both conflicts capitalized on local resources as much as they could, the people providing them information and some direct support in the forms of guiding, tracking, scouting, and labor. The jungle provided water, cover, concealment, and food. The development of good British marksmanship in both struggles was also essential to success. Tactical accounts of the campaigns repeatedly stress this subject. Winning and surviving for infantrymen in Borneo and Malaya meant shooting first, shooting straight, and shooting to kill.

At the strategic level, the Briggs Plan and the implementation of General Walker's six principles, both of which established the framework for a successful, integrated approach to the two wars, must be given great credit. Both plans assumed a long-term national commitment would be necessary to achieve a solution. The British acknowledged from the outset that ridding the country of the threat would require years of sacrifice. An impulsive or over-hasty approach would never have worked. This attitude of perseverance seeped down to the tactical level, where the leaders and their men accepted long-term personal commitments.

Implicit in the Briggs Plan and in Walker's principles was the idea that in these kinds of wars, military operations must yield to political authority. The strategic leadership understood that achieving military victory was meaningless if it was divorced from political objectives. Thus, the directors of operations accepted what may have seemed to be illogical restrictions on their use of force, generally unaccompanied by the gnashing of teeth that was commonplace during the Vietnam War. While the directors certainly argued their own points of view, they understood that low-intensity warfare always involves politically induced, artificial restrictions on operations. They recognized

144

that in such situations, political actions were paramount, and military actions were only supportive. Ultimately, the military forces would leave; the political structure left behind had to be able to stand by itself.

The joint systems for directing the war effort worked well in both wars, but problems existed. In Malaya, the DWEC or SWEC solution to a problem seldom was the optimal solution. The products of compromise, DWEC and SWEC decisions were prompted first by an attempt to achieve unity among the various civil, military, and police agencies and only secondly to effect military efficiency. In Borneo, Walker and his subordinates had to contend with several police forces and civil bureaucracies that they did not control. To have their way, they had to persuade, convince, and cajole. Still, the principle that the civil, police, and military organizations had to operate hand in hand in harmony was observed. Cooperation, not competition, was the goal. Within the military services themselves, it was equally important to develop unity by having one man direct the air, army, and naval forces. In the words of one battalion commander: "If nothing else was learned in the Borneo campaign except the necessity to have a national, flexible joint organization to fight a common battle, then the three years along the border will not have been wasted."[54] Many others had the same thing to say about the Emergency.

The strategic leadership also recognized that the best ideas on tactical operations came from the units themselves. Directors of operations put their staffs to work collecting and evaluating these ideas, not dreaming up their own notions and foisting them on the infantry units.

The British approach to winning hearts and minds deserves a great deal of study in the U.S. This principle was vital to success at every level and was impressed on the lowest soldier in the chain of command. For infantrymen, it basically meant protecting the citizens of the country, respecting their persons and their property, and not being afraid to get close to them.

Commonwealth and friendly nation students at the Jungle Warfare School

145

Finally, the British Army showed great wisdom in their appointments of generals to high positions in the two wars. First Briggs, then Templer, placed Malaya squarely on the road to success against the Communist terrorists. From the precedence they set, their successors knew that their best course was to steer straight ahead. In Borneo, the extremely capable Walker was replaced by another officer of high ability, Major General George Lea, former commander of the 22d SAS. These men were uniquely qualified to carry out their duties wisely and efficiently. Tactically experienced and gifted with strategic vision, they were able, by virtue of their forceful leadership, to dynamically combine small light infantry actions with the overall war strategy.

UNCLASSIFIED

THE CONDUCT

OF

ANTI-TERRORIST OPERATIONS

IN

MALAYA

(Third Edition—1958)

> The information given in this
> document is not to be communicated,
> either directly or indirectly, to the
> Press or to any person not authorised
> to receive it.

FOR OFFICIAL USE ONLY

UNCLASSIFIED

*Source: Great Britain, Army, Director of Operations, Malaya. *The Conduct of Anti-Terrorist Operations in Malaya*, 3d ed. (1958), i—ix.

MO...........RIZED

PART 1
GENERAL

CHAPTER I
MALAYA

CHAPTER II
THE CT—ORIGIN AND DEVELOPMENT

MODIF............RIZED

ii

4

CHAPTER III
OWN FORCES

CHAPTER IV
THE EMERGENCY REGULATIONS (ERs) AND METHODS OF SEARCHING

APPENDIX

PART 2
OPERATIONS
CHAPTER V
PLATOON ORGANISATION, WEAPONS AND EQUIPMENT

APPENDIX

CHAPTER VI

THE JUNGLE BASE

CHAPTER VII

PATROLLING

CHAPTER VIII

PATROL MOVEMENT AND FORMATION

UNCLASSIFIED

7

UNCLASSIFIED

CHAPTER XV

TRAINING FOR OPERATIONS

CHAPTER XVI

WIRELESS COMMUNICATIONS IN MALAYA

CHAPTER XVII

AIR SUPPORT IN MALAYA

APPENDICES

CHAPTER XVIII

EMPLOYMENT OF THE ROYAL ARTILLERY IN ANTI-CT OPERATIONS

10

CHAPTER XIX

HANDLING OF ABORIGINES BY SECURITY FORCES

CHAPTER XX

THE EMPLOYMENT OF DOGS ON OPERATIONS AND THE ADMINISTRATION OF WAR DOGS

CHAPTER XXI

TRACKING

ll

PART 3

ADMINISTRATION

CHAPTER XXII

OPERATIONAL RATIONS

CHAPTER XXIII

FIRST AID AND PREVENTIVE MEDICINE

Appendix B*
Patrol and Ambush Orders

108

Appendix A
PATROL ORDERS—AIDE MEMOIRE

1. **Situation**

   (a) Topography.—Use maps, air photos, visual recce and patrol going map.

   (b) CT in Area:—
       (1) Strength.
       (2) Weapons and dress.
       (3) Known or likely locations and activities including past history.

   (c) Movements of Aborigines and civilians in area.

   (d) Own troops:—
       (1) Clearance.
       (2) Patrol activities of SF. Include means of identification.
       (3) Air and arty tasks

2. **Mission**

   This must be clear to patrol commander:—

   (a) Recce Patrol.—takes form of question or questions.

   (b) Fighting Patrol.—definite object.

3. **Execution**

   (a) Strength and composition of patrol.

   (b) Time out and anticipated time of return.

   (c) Method of movement to patrol area.

   (d) Routes out and in. If helicopters are to be used location and state of LZs.

   (e) Boundaries.

   (f) Probable bounds and RVs.

   (g) Formations.

   (h) Deception and cover plan.

   (j) Action to be taken on contact.

   (k) Action if ambushed.

   (l) Action if lost.

   (m) DO NOT:—
       (1) Move in file through rubber.
       (2) Move through defiles.
       (3) Cut unnecessarily.
       (4) Return by the same route as that used for outward move.
       (5) Allow weapons to become jammed through dirt.
       (6) Relax because you are nearing base.

4. **Administration and Logistics**

   (a) Rations:—
       (1) Type and number of days.
       (2) Resupply.

*Source: Great Britain, Army, Director of Operations, Malaya. *The Conduct of Anti-Terrorist Operations in Malaya*, 3d ed. (1958), XI.

109

      (3) Cooking.
      (4) Dog rations.
      (5) Rum.
  (b) Equipment and Dress:—
      (1) Change of clothing.
      (2) Large or small pack.
      (3) Poncho capes.
      (4) Footwear.
      (5) Maps, compasses, and air photos.
  (c) Avoidance of noise:—
      (1) Does equipment rattle?
      (2) Leave behind men with coughs.
  (d) Weapons:—
      (1) Types and distribution.
      (2) Special weapons—GF rifle, etc.
  (e) Ammunition:—
      (1) Type and distribution.
      (2) Grenades, Hand and Rifle, including gas checks and clips for 80 grenades.
      (3) Check grenade fuzes.
      (4) Signal cartridges.
  (f) Medical:—
      (1) First field dressing, J packs.
      (2) Medical orderly and haversack.
      (3) Water sterilising tablets.
      (4) Salt tablets.
      (5) Paludrine.
      (6) DBP clothing.
      (7) Foot powder.
      (8) Copper sulphate ointment for burns.
  (g) Special Equipment:—
      (1) Saws and parangs.
      (2) Cameras.
      (3) Finger print outfit.
      (4) Surrender pamphlets.
      (5) Night equipment.
      (6) Explosives.
      (7) Dogs.
      (8) Marker Balloons.
  (h) Inspect all equipment for serviceability.

3. **Command and Signals**

  (a) Frequencies:—
      (1) Times of opening.
      (2) Special instructions.
      (3) Air.

110

(b) Codes:—
  (1) Net identification signs.
  (2) Codes.
  (3) Passwords.

(c) Check and test sets:—
  (1) Aerials.
  (2) CW keys.
  (3) Spare batteries.

(d) Ground/Air Communications:—
  (1) DZ panels and DZ letters allotted.
  (2) Ground/Air signal code.

NOTE:

Check thoroughly that all points have been understood by patrol members.

Appendix A

AMBUSH ORDERS—AIDE MEMOIRE

REMEMBER SECURITY—DO NOT USE THE TELEPHONE DO NOT ALLOW MEN OUT AFTER BRIEFING

Suggested Headings

1 **Situation**

    (a) *Topography*.—Use of air Photographs, maps and local knowledge consider use of a guide.

    (b) *CT*.—

        (1) Expected strength.
        (2) Names and anticipated order of march. Photographs.
        (3) Dress and weapons of individuals.
        (4) Which is the VIP.
        (5) What are habits of party concerned.

    (c) *SF*.—

        (1) Guides or SEP to accompany.
        (2) What other SF are doing.

    (d) *Clearance*.—

        (1) Challenge.
        (2) Password.
        (3) Identifications.

    (e) *Civilians*.—

        (1) Locations.
        (2) Habits.

2. **Mission**

    This must be clear in the mind of every man especially when a particular CT is to be killed.

3. **Execution**

    (a) Type of layout.
    (b) Position and direction of fire of groups.
    (c) Dispersal point.
    (d) Weapons to be carried.
    (e) Composition of groups.
    (f) Timings and routes.
    (g) Formations during move in.
    (h) Orders re springing.

(j) Distribution of fire.

(k) Use of grenades.

(l) Action on ambush being discovered.

(m) Order to cease firing.

(n) Orders re immediate follow up.

(o) Orders for search.

(p) Deliberate follow up.

(q) Signal to call off ambush.

(r) Rendezvous.

(s) Dogs—if any.

(t) Deception plan.

(u) Alerting.

4. Administration and Logistics

(a) Use of transport to area.

(b) Equipment and dress:—
    Footwear for move in

(c) Rations—if any.

(d) Special equipment:—
    (1) Night lighting equipment.
    (2) Cameras.
    (3) Finger print equipment.

(e) Medical:—
    (1) First field dressing, first aid packs.
    (2) Medical Orderly.
    (3) Stretcher and ambulance.

(f) Reliefs.

(g) Administrative Area, if required orders re cooking, smoking

(h) Transport for return journey.

(j) Inspection of personnel and equipment:—
    (1) Men with colds not to be taken.
    (2) Is zeroing of weapons correct?
    (3) Is ammunition fresh?
    (4) Are magazines properly filled?

5. Command and Signals

Success signal.

NOTES

Chapter 3

1. Great Britain, Army, Director of Operations, Malaya, *The Conduct of Anti-Terrorist Operations in Malaya*, 3d ed. (1958), II—6, hereafter cited as the *ATOM Manual*.

2. Richard Miers, *Shoot to Kill* (London: Faber and Faber, 1959), 29.

3. *ATOM Manual*, III—5; J. B. Oldfield, *The Green Howards in Malaya* (Aldershot, Hampshire, England: Gale and Polden, 1953), 21—22; and Richard L. Clutterbuck, *The Long, Long War: Counterinsurgency in Malaya and Vietnam* (New York: Frederick A. Praeger, 1966), 60—64.

4. Jac Weller, *Fire and Movement: Barguin-Basement Warfare in the Far East* (New York: Thomas Y. Crowell Co., 1967), 46. This estimate may be high. Clutterbuck, *Long, Long War*, 60, notes that 423,000 persons were resettled in one year.

5. Miers, *Shoot to Kill*, 29; and M. C. A. Henniker, *Red Shadow over Malaya* (London: William Blackwood and Sons, 1955), 50.

6. Henniker, *Red Shadow*, 297.

7. Miers, *Shoot to Kill*, 82.

8. Henniker, *Red Shadow*, 297.

9. Miers, *Shoot to Kill*, 29.

10. Henniker, *Red Shadow*, 16.

11. Oldfield, *Green Howards*, 5; and Clutterbuck, *Long, Long War*, 51.

12. Henniker, *Red Shadow*, 16.

13. Clutterbuck, *Long, Long War*, 50.

14. Ibid., 100.

15. Oldfield, *Green Howards*, 33.

16. Clutterbuck, *Long, Long War*, 52.

17. Miers, *Shoot to Kill*, 52.

18. Ibid., 137—98.

19. *ATOM Manual*, V—1.

20. Ibid., V—2.

21. Ibid., V—7.

22. Ibid., V—3 through V—6.

23. The stress given to quick, accurate fire by the cadre at the JWS and by unit leaders cannot be overemphasized.

24. Tony Geraghty, *Inside the Special Air Service* (Nashville, TN: Battery Press, 1981), 33. One estimate calculated that 1,800 man-hours in patrolling were expended for every contact. The ratio of man-hours employed to kill was much higher of course.

25. Oldfield, *Green Howards*, xxiii.

26. *ATOM Manual*, chapter VIII.

27. Geraghty, *Special Air Service*, 22.

28. Miers, *Shoot to Kill*, 103.

163

29. Oldfield, *Green Howards*, 28. An example cited here describes a platoon patrol which covered sixty miles in thirteen days.

30. *ATOM Manual*, chapter XI.

31. Ibid., XI—2.

32. Oldfield, *Green Howards*, 98.

33. Miers, *Shoot to Kill*, 201.

34. Tom Pocock, *Fighting General: The Public and Private Campaigns of General Sir Walter Walker* (London: Collins, 1973), 104.

35. *ATOM Manual*, XI—15.

36. Henniker, *Red Shadow*, 166.

37. Ibid., 249, 260.

38. Oldfield, *Green Howards*, 63; and Geraghty, *Special Air Service*, 34.

39. Henniker, *Red Shadow*, 166.

40. Harold Douglas James and Denis Sheil-Small, *The Gurkhas* (Harrisburg. PA: Strokpole Books, 1966), 262.

41. Miers, *Shoot to Kill*, 160.

42. Henniker, *Red Shadow*, 133—35.

43. *ATOM Manual*, chapter XIII.

44. Henniker, *Red Shadow*, 256. The author relates how a Fijian infantry unit known for its excellence on patrol violated the route SOP one day and lost six wounded and five killed to a CT ambush.

45. Ibid., 50.

46. J. A. C. Mackie, *KONFRONTASI: The Indonesia-Malaysia Dispute, 1963—1966* (New York: Oxford University Press, 1974), 215.

47. Pocock, *Walter Walker*, 201, 215.

48. Harold Douglas James and Denis Sheil-Small, *The Undeclared War: The Story of the Indonesian Confrontation, 1962—1966* (Totowa, NJ: Rowman and Littlefield, 1971), 56.

49. David Lee Watkins, "Confrontation: The Struggle for Northern Borneo," MMAS thesis, U.S. Army Command and General Staff College, 1978, 25.

50. "Jungle Patrol, 2nd Battalion the Parachute Regiment in Borneo," *The Infantryman* 81 (November 1965):74—75, reprinted from *Pegasus*, the journal of the British Airborne Forces, hereafter cited as "Jungle Patrol." See also Lieutenant Colonel A. S. Harvey, "Random Reflections of an Infantry Battalion Commander on the Indonesian Border," *The Infantryman* 81 (November 1965):50—51.

51. Pocock, *Walter Walker*, 93.

52. Ibid., 137.

53. Ibid., 159—60.

54. Ibid., 172.

55. James and Sheil-Small, *Undeclared War*, 99—101.

56. Peter Dickens, *SAS, the Jungle Frontier: 22 Special Air Service Regiment in the Borneo Campaign, 1963—1966* (London: Arms and Armour Press, 1985), 65.

57. Lieutenant Colonel P. E. B. Badger, "Tigers in the Jungle," *The Infantryman* 80 (October 1964):50.

58. "Jungle Patrol," 73. Planning considerations and operational parameters for the conduct of helicopter operations in Borneo are discussed in Air Vice Marshal C. N. Foxley-Norris, "Air Aspects of Operations Against 'Confrontation'," in *Brassey's Annual: The Armed Forces Yearbook, 1967*, edited by Major General J. L. Moulton, et al. (New York: Frederick A. Praeger, 1967), 281—91.

59. Dickens, *SAS*, 88.

60. "Jungle Patrol," 74—75.

61. James and Sheil-Small, *Undeclared War*, 86.

62. Pocock, *Walter Walker*, 168.

63. Weller, *Fire and Movement*, 48.

64. Lieutenant Colonel H. J. Sweeney, "Long Range Patrolling," *The Infantryman* 81 (November 1965):20.

65. Geraghty, *Special Air Service*, 46.

66. Ibid., 66. In Malaya, the SAS worked in the deep jungle with the tribal aborigines. They functioned there primarily in the role of "Hearts and Minds." They also provided terrain reconnaissance, set booby traps on CT camps and food caches, conducted river patrols, mounted ambushes, and experimented in deep insertion through parachuting into the jungle canopy (tree jumping). See ibid., 26—32.

67. Dickens, *SAS*, 53.

68. Ibid., 71.

69. Ibid., 75.

70. Ibid., 102.

71. Ibid., 34. It should also be noted that due to a shortage in SAS personnel, British and Gurkha paratroop companies were converted to SAS squadrons.

72. Pocock, *Walter Walker*, 197.

73. Ibid.

74. Geraghty, *Special Air Service*, 57.

75. James and Sheil-Small, *Undeclared War*, 170.

76. Watkins, "Confrontation," 90.

77. Ibid., 81.

78. Dickens, *SAS*, 36.

79. Pocock, *Walter Walker*, 192.

80. Weller, *Fire and Movement*, 51—56, contains an excellent discussion of the equipment and weaponry issues.

81. Ibid., 56; and Pocock, *Walter Walker*, 192.

82. Pocock, *Walter Walker*, 216.

83. Ibid., 189. See also, Watkins, "Confrontation," 87.

84. Quoted in Watkins, "Confrontation," 73.

Chapter 3

Badger, P. E. B., Lieutenant Colonel. "Tigers in the Jungle." *The Infantryman* 80 (October 1964):48—56.

Clutterbuck, Richard I. *The Long, Long War: Counterinsurgency in Malaya and Vietnam.* New York: Frederick A. Praeger, 1966.

Dickens, Peter. *SAS, the Jungle Frontier: 22 Special Air Service Regiment in the Borneo Campaign, 1963—1966.* London: Arms and Armour Press, 1985.

Foxley-Norris, C. N., Air Vice Marshal. "Air Aspects of Operations Against 'Confrontation.' " In *Brassey's Annual: The Armed Forces Yearbook, 1967,* edited by Major General J. L. Moulton, et al. New York: Frederick A. Praeger, 1967.

Geraghty, Tony. *Inside the Special Air Service.* Nashville, TN: Battery Press, 1981. Originally published in England under the title *Who Dares Win* (London: Arms and Armour Press, 1980).

Great Britain. Army. Director of Operations, Malaya. *The Conduct of Anti-Terrorist Operations in Malaya.* 3d ed. 1958.

Hall, P. G. "The Jungle Warfare School." *The Infantryman* 82 (November 1966):45—49.

Harvey, A. S., Lieutenant Colonel. "Random Reflections of an Infantry Battalion Commander on the Indonesian Border." *The Infantryman* 81 (November 1965):47—51.

Henniker, M. C. A. *Red Shadow Over Malaya.* London: William Blackwood and Sons, 1955.

James, Harold Douglas, and Denis Sheil-Small. *The Gurkhas.* Harrisburg, PA: Stackpole Books, 1966.

———. *The Undeclared War: The Story of the Indonesian Confrontation, 1962—1966.* Totowa, NJ: Rowman and Littlefield, 1971.

"Jungle Patrol, 2nd Battalion the Parachute Regiment in Borneo." *The Infantryman* 81 (November 1965):73—75. Reprinted from *Pegasus,* the journal of the British Airborne Forces.

Mackie, J. A. C. *KONFRONTASI: The Indonesia-Malaysia Dispute, 1963—1966.* New York: Oxford University Press, 1974.

Miers, Richard. *Shoot to Kill.* London: Faber and Faber, 1959.

Oldfield, J. B. *The Green Howards in Malaya*. Aldershot, Hampshire, England: Gale and Polden, 1953.

Pocock, Tom. *Fighting General: The Public and Private Campaigns of General Sir Walter Walker*. London: Collins, 1973.

Sweeney, H. J., Lieutenant Colonel. "Long Range Patrolling." *The Infantryman* 81 (November 1965):19—22.

Warner, Philip. *The Special Air Service*. London: William Kimber, 1971.

Watkins, David Lee. "Confrontation: The Struggle for Northern Borneo." MMAS thesis, U.S. Army Command and General Staff College, Fort Leavenworth, KS, 1978.

Weller, Jac. *Fire and Movement: Bargain-Basement Warfare in the Far East*. New York: Thomas Y. Crowell Co., 1967.

Chapter 4

The First Special Service Force

Introduction

The proliferation of specially organized and trained units among the Allied armies in World War II assumed astounding proportions. Airborne divisions, commando units, Ranger battalions, Special Operations Executive and Office of Strategic Services detachments, Marine raiders: special units of all sorts abounded. This study has already examined one of those specialty forces, the Chindits. This chapter will introduce another force, a unique Canadian-American brigade called the First Special Service Force (FSSF, also called the Force). Raised, manned, and trained as a light infantry raiding force to be used in cold, high mountains, the FSSF established a well-deserved record as one of the toughest and most effective combat units in Europe. Because of its uniqueness, however, the Force should not be considered a prototype for modern light infantry forces. While its training, tactical techniques, and operational record merit study, the relevant lessons are both positive and negative in nature; there is much to be emulated and much to be avoided in its example.

Geoffrey Pyke, an eccentric British scientist, originated the idea that led to the creation of the FSSF. Pyke convinced Prime Minister Winston Churchill and Chief of Combined Operations Lord Mountbatten that the Allies needed to develop a light over-snow vehicle that could then be used as the primary transport for mobile raiding forces sent on strategic missions of sabotage in remote areas. In particular, such a vehicle and force could be used to knock out the important hydroelectric stations in German-occupied Norway. Through Churchill's influence, the United States agreed to build the vehicles, while the United States and Canada jointly agreed to supply the men for the Force. For various reasons, Operation Plough, as it came to be known, never was mounted, although the vehicle (the M-24 Weasel) was developed and used by the 10th Mountain Division and other units in northern Italy.[1] The Canadian-American force established for Operation Plough narrowly survived the cancellation of the Norwegian operation, and for some time, no one could decide where to use it.[2] Throughout the early period of its formation, the FSSF retained its intended missions of raiding, sabotage, or spearhead operations in cold, mountainous regions.

Selection and Organization

Once the decision to establish the FSSF was made, both sponsoring nations issued a call for volunteers. American Lieutenant Colonel Robert T. Frederick, named as the Force commander, requested the assignment to his unit of "single men between ages of 21 and 35 who had completed three years or more grammar school within the occupational range of Lumberjacks, Forest Rangers, Hunters, North woodsmen, Game Wardens, Prospectors, and Explorers."[3] The Canadian Army established more exacting standards. It said its volunteers must:

1. Be willing to undergo airborne training.

2. Be physically fit.

3. Be already fully trained as infantrymen.

4. Possess a knowledge of internal combustion engines (in anticipation of driving and maintaining the Weasel).

5. Be NCO (noncommissioned officer) material, since the standard enlisted rank in the Force was to be sergeant.

6. Have experience as mountaineers, skiers, or woodsmen, or have had winter training.[4]

From the start, the Canadian cohort comprised soldiers of higher quality and motivation than the average soldier. However, the initial American component included a large percentage of jailbirds, ne'er-do-wells, and other culls—as unit and post commanders in the United States took advantage of the call for volunteers to rid themselves of their troublemakers. Although a great many of these doubtful recruits were turned away, a substantial "disreputable" element remained. Those that were accepted into the unit possessed the rugged and somewhat reckless character sought by Frederick. Eventually Frederick assembled a force of individualistic, tough soldiers who were ready to be molded into a fighting arm, steadied by the influence of their more disciplined and initially better-trained Canadian comrades. Interestingly, the average age of the men of the Force during its first year of organization and training was twenty-six—an age considerably higher than that of men in regular units. The Force's executive officer, Lieutenant Colonel Paul Adams, attributed the strong unit cohesion and maturity of the Force to this older average age.[5]

The original concept for the employment of the FSSF in Norway called for 18 companies of 100 men each. As a result, Frederick organized the unit into 3 small regiments of approximately 600 men each. Each regiment was composed of two battalions, each with three companies (see figure 13). The companies included three platoons, each with two twelve-man sections led by staff sergeants. The three regiments comprised the combat echelon of the Force. All support functions were performed by the separate 600-man, all-American Service Battalion.

The Service Battalion was an experiment of sorts. It was created to relieve the combat echelon of any noncombat duties that might detract from its training or operations. The Service Battalion was divided into three companies. The headquarters company included the Force headquarters, clerks, air detachment, communications detachment, and a military police platoon. The maintenance company performed all vehicle and weapons maintenance. The service company provided cooks, bakers, riggers, barbers, supply sergeants, and porters to support the Force. Finally, the medical detachment, headed by the Force's surgeon, provided medics and operated the unit aid stations.[6] Frederick appeared to be satisfied with this initial organization of the FSSF into discrete combat and service echelons. The unit retained this basic form throughout its history.

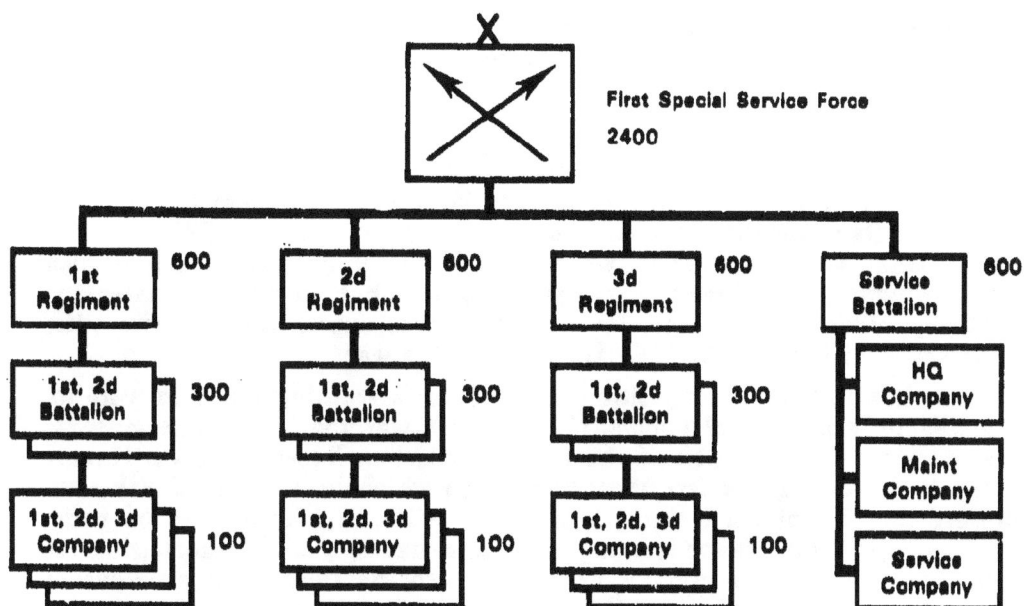

Note: Figures above are approximate manpower. Companies included three platoons, each with two 12—16 man sections.

Figure 13. FSSF organization

The official U.S. Army history of World War II states that the FSSF was authorized 1,190 trucks and cars and 600 T-24 Weasels.[7] It did not, however, receive vehicles in such numbers. For instance, it received only small numbers of Weasels, and these on a temporary basis. Throughout its operations in Italy, the Force periodically scrounged its needed transport.

Initially only lightly armed, the Force's table of organization and equipment (TOE) was changed when the Norway operation was canceled. This move was based on the correct assumption that the FSSF would be used in more conventional operations. To ensure that his regiments would conduct sustained operations, Frederick supplied each section with one Browning automatic rifle (BAR), one Johnson light machine gun (a Marine weapon that the men of the Force liked better than the BAR because it was lighter and could be operated by one man), and a bazooka or a 60-mm mortar. Most infantrymen carried M-1 rifles. Officers carried carbines or pistols. Section leaders carried Thompson submachine guns. The Force lacked organic artillery, heavy mortars, medium-heavy machine guns, and armor of any sort. The Johnson light machine guns were not authorized: Frederick traded 2 tons of a new explosive (RS) to the Marines Corps in exchange for 125 of the weapons.[8] Some sections also possessed flamethrowers.

Except for the Service Battalion, Canadians and Americans were mixed evenly throughout the Force. Although the Canadians only numbered about 600, or one-third of the combat echelon, they occupied about one-half of the leadership positions. The Canadian soldiers were generally older and more

171

experienced than the Americans, so most of the battalion commanders and senior sergeants were Canadians, while most of the junior officers were Americans.

Training

Frederick selected Fort William Henry Harrison, an unused training site in the Montana mountains near Helena, as the base for the FSSF. As the many Canadian and American contingents arrived in July 1942 from all over North America, they began a training program that, in terms of intensity, difficulty, variety, and scope, far surpassed that experienced by any other regiment or division in the U.S. Army during the war.

The intensity and composition of the Force's training program were influenced by a number of factors. One of these was the target date given to Frederick for the execution of the Norway mission, December 1942. Having much to do and only a few short months in which to do it, Frederick compressed the time of the training wherever feasible, and he insisted on using every available minute for training short of exhausting and demoralizing his command. Thus, Force troopers underwent airborne training in six days as opposed to the normal three weeks. The daily training day normally extended from 0430 to 1700. However, during four evenings a week, lectures on various subjects were presented from 1900 to 2100. Generally, the men of the Force enjoyed only Saturday nights and Sundays off.

The Force's training was tailored to prepare the men for operations in cold weather and mountainous regions. To ready themselves for this environment, members of the Force developed special skills such as skiing with pack and rifle, rock climbing, mountain walking, use of ropes, survival, and the operation and maintenance of snow carriers. The Force spent many days and nights in the Montana mountains learning how to cope, move, and fight under frigid conditions in very rugged ground. The Norwegian experts that Frederick imported to train his command were amazed at how rapidly and proficiently his soldiers learned their needed mountain skills.[9] As part of the training, conditions were made austere. The men, for example, lived in railroad boxcars while participating in ski training near Blossburg, Montana.[10]

Frederick established two baseline goals for individual training: each soldier was to reach an unusually high level of physical fitness and stamina, and each soldier was to be a consummate infantryman. These two requirements formed the foundation upon which unit capability was built.

The physical-training program was extraordinarily difficult for its time. An official Canadian report described the physical training in this manner:

> The programme of physical training was designed to produce a standard of general fitness and stamina capable of meeting the severest demands made upon it by fatigue of combat, unfavorable terrain or adverse weather.
>
> This physical training has been built up to such a pitch that an ordinary person would drop from sheer exhaustion in its early stages. This course includes crawling, rope-climbing, boxing, push-ups, games, much doubling and running.[11]

The men routinely double-timed from one training site to the next, and they ran the 1 1/2-mile obstacle course as a daily requirement. Long marches at exaggerated paces with heavier-than-normal loads occurred frequently. At one point, the S3 laid out a sixty-mile course to see which regiment could complete it most rapidly. The 1st Regiment came in after twenty hours of dogged marching with full packs.[12] A strong spirit of rivalry characterized such events. All men of the Force, including officers, had to meet the standard, an impossible task for the aged and unfit. Those men failing to measure up received little mercy or compassion. They were dropped from the rolls and returned to their original units. Later, when the Force was committed to operations, its commanders realized that this high level of fitness deteriorated gradually in combat. Consequently, units always used their rest periods out of the line to renew physical strength and stamina through exercise.

Frederick was just as insistent that soldiers master a wide range of infantry skills. Foremost among these were marksmanship. Regiments constantly had groups at work qualifying on all the unit's small arms, including automatic weapons, bazookas, mortars, and flamethrowers. Training was also conducted on captured German weapons.[13]

Training also emphasized demolitions because of the intended use of the Force for raids and sabotage. Experts, again, were brought in to train the men, and some of the officers attended a two-week course in demolitions that qualified them as trainers. Men of the Force seemed to take to demolitions naturally, and they showed great delight in the training. Occasionally, an overexuberant flair led the men to use too much explosive material on targets, which blew out windows in nearby towns. Paul Adams explained this penchant: "We decided we wanted to be sure to never have to go back and try it again."[14]

Parachute training and hand-to-hand combat served to cement an attitude of recklessness, daring, and aggressiveness within the Force. Anyone refusing to don a parachute and jump out of an airplane got a train ticket home the next day. Frederick obtained an expert, Irishman Pat O'Neill, to teach his men a mixture of karate, jujitsu, and trick fighting. Again, the soldiers turned to this training with a certain élan, trying out their new skills on each other, local miners, lumberjacks, or MPs. Bayonet training proceeded with bare blades and included the officers. Adams, the Force executive officer, recalls how he was nicked in the neck by Frederick, an extraordinarily fit and agile man, when the two were paired for bayonet drills.[15]

Infantry training also included the standard subjects of first-aid, camouflage, scouting and patrolling, map reading, and unit tactics. Some attention was also given to irregular tactics. This training was not significantly different from regular infantry training except for its adaptation to mountain terrain and its emphasis on raiding.[16]

As the FSSF became proficient in its training, more and more of it was conducted at night. For example, the unit conducted only two lessons on land navigation in daylight; the rest of the training took place at night.[17] The Force leadership was fully convinced that investment in night operations in training would pay big dividends in combat.

These ingredients formed the recipe for Force training from July to December 1942. The intensity, difficulty, and stressful nature of the training produced a strong bond of cohesion and esprit within the Force and a powerful sense of confidence and derring-do among those who were not eliminated. In such an environment, the Force quickly jelled into an effective elite unit. National distinctions disappeared as the men found their identity as members of the Force.

The command peaked in its training in early December, but by then, it had nowhere to go. To avoid staleness, Frederick put everyone on leave who wanted to go—in staggered increments. January and part of February were then given over to regimental maneuvers that stressed sabotage operations behind enemy lines against specific objectives, such as tunnels, bridges, and dams. The Force then turned to company and battalion exercises during the remainder of February and March.

In April, the FSSF moved to Camp Bradford near Norfolk, Virginia, for amphibious training. The Force still had no stated area for deployment, although its use in the invasion force being raised to retake the Aleutians from the Japanese had been discussed. At Camp Bradford, the Force raced through the amphibious training, completing the basic instruction a week ahead of schedule. Simulated combat landings followed for another seven to ten days. The report filed by the Amphibious School staff highly praised the FSSF and stated that it was fully qualified for any amphibious operations.[18]

Next, the FSSF arrived at Fort Ethan Allen in Vermont, where it conducted additional training in landings by rubber boat, scouting, patrolling, raiding, and demolitions. Finally, the War Department directed that the Force be used in the North Pacific against Kiska Island. Shortly thereafter, Army Ground Forces sent out an inspection team to certify the Force's readiness:

A score of Corps inspectors hit Fort Ethan Allen on June 15, armed with the latest Ground Forces standards of training, equipment, and physical fitness for troops bound for the fighting fronts. All Force units were to undertake the various tests and be rated. Perfection was one hundred percent on the rated test but a unit could pass with a seventy-five percent average. In the first test, each company had to cover a four-mile march route under rifle and pack in one hour's time, points to be deducted for overtime and stragglers on the march. Other tests covered proficiency in map reading, maintenance of weapons, the usual military subjects by oral quizzes, and still other examinations covered physical fitness, calisthenics and foot races. At no point did the established standards adequately rate the Force units, who scored on an average about 125 percent (on some tests, 200 percent), and drew the inspectors' admissions that the standard test structure could not be applied to a unit as well trained as the First Special Service Force. They expressed amazement several times at the loads and quantities of ammunition the men carried easily. Another thing that greatly impressed the inspectors was the absolute and thorough coordination within the elements of the Force, and the complete confidence each man had in himself and his comrades. They expressed great satisfaction in the realism injected into Force training.

The Corps group also watched the company demolition problems, took a look at the live-firing range—the battle orientation course—and sailed out on Lake Champlain to view the intelligence scouts in rubber-boat and cliff-scaling exercises. Then Army Ground Forces was notified: Yes, the First Special Service Force was ready for any job that had to be done.[19]

174

Operations

The operational history of the FSSF began on the early morning of 15 August 1943, when the 1st and 2d Regiments silently waded ashore on Kiska Island in the Aleutian chain under cover of darkness. This operation had been carefully planned and practiced. Furthermore, the FSSF had been assigned a mission appropriate to its capability, that of securing two separate beachheads on the island in advance of the landing of the invasion's main forces. The 3d Regiment was held in airborne reserve (see map 12).

The mission was fraught with danger. It demanded endurance and great skill in the handling and control of small rubber boats, which would be paddled ashore in the dark, in very cold water. The mission also required exceptional stealth and silence to achieve surprise, which would be followed by the likelihood of a fierce, close-in battle against a tough enemy that would initially

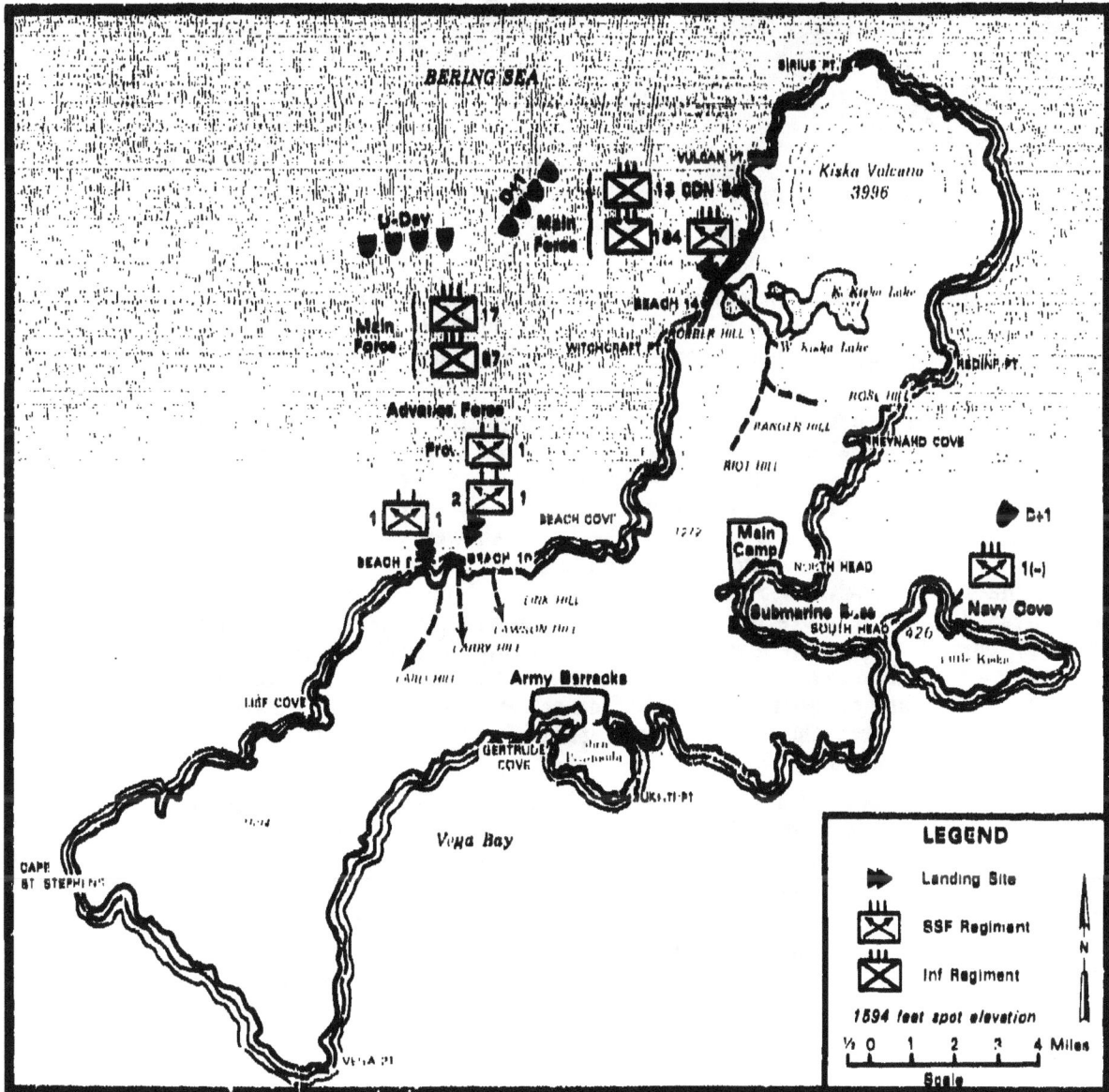

Source: Burhans, *The First Special Service Force,* 73.

Map 12 Landings on Kiska Island, 15—16 August 1943

Weasels, bulldozers, and Athey trailers on Lilly Beach, Kiska Island

outnumber the Force. In addition, the weather was harsh and the terrain difficult. In short, this was a mission perfectly suited to a unit like the Force, and Commanding General Simon B. Buckner was wise to have assigned it to them.

As it turned out, the Japanese had evacuated the island days earlier. The hard fight anticipated by the planners turned out to be an uncontested occupation. Even so, the operation had great value as a training exercise and permitted the Force to display a number of the characteristics that came to be associated with it. For instance, the Force showed outstanding technical ability in controlling the amphibious landing and in coordinating its advance in the dark to positions overlooking the beaches. Moreover, no straggling occurred during the move inland across the rough tundra, even though each man carried an average load of 90 to 100 pounds. In addition, fire discipline was very good compared with that exhibited by the troops that came ashore later. All tasks had been accomplished on schedule with elan and steady confidence. These strengths were cited in a letter of commendation from the army task force commander.[20]

Unblooded but not untested, the FSSF left the North Pacific promptly, en route eventually to Europe. General Eisenhower had requisitioned the unit for service in the Mediterranean, where he envisioned its use for special reconnaissance, for raids behind German lines in Italy, or for independent guerrilla operations and sabotage in the Balkans to support resistance groups.

The Force never made it to the Balkans but was destined to spend quite some time in Italy in a variety of roles, several of them inconsistent with its design and training. Assigned to the Fifth Army from November 1943 to June

176

1944, the FSSF was immediately committed to the assault of a seemingly inpregnable German stronghold atop Mount de la Difensa. Thereafter, it found itself employed in succession as a standard, separate infantry brigade in the line; as a unit employed in flank protection in the high mountains that paralleled the route of advance of main forces; as a beachhead force at Anzio, holding the line against the Germans for over three months; as a spearhead force used to break out of the beachhead; again as flank protection; and finally as the infantry component of the tank-infantry spearhead leading the Fifth Army into Rome. Each of these operations will be discussed in turn, but only the attack against Difensa and activities at Anzio will be covered in detail, since they are the most instructive.

Operation Raincoat

Mount de la Difensa was part of a large, high complex of peaks and ridgelines known as the Camino hill mass. Considered to be vital terrain by both sides, the mountains rose precipitously on the south side of the Mignano Gap, a narrow valley that opened into the large Liri Valley, the "Gateway to Rome." Running roughly six miles long by four miles wide, the Camino hill mass averaged about 3,000 feet in height (see map 13). The slopes facing the

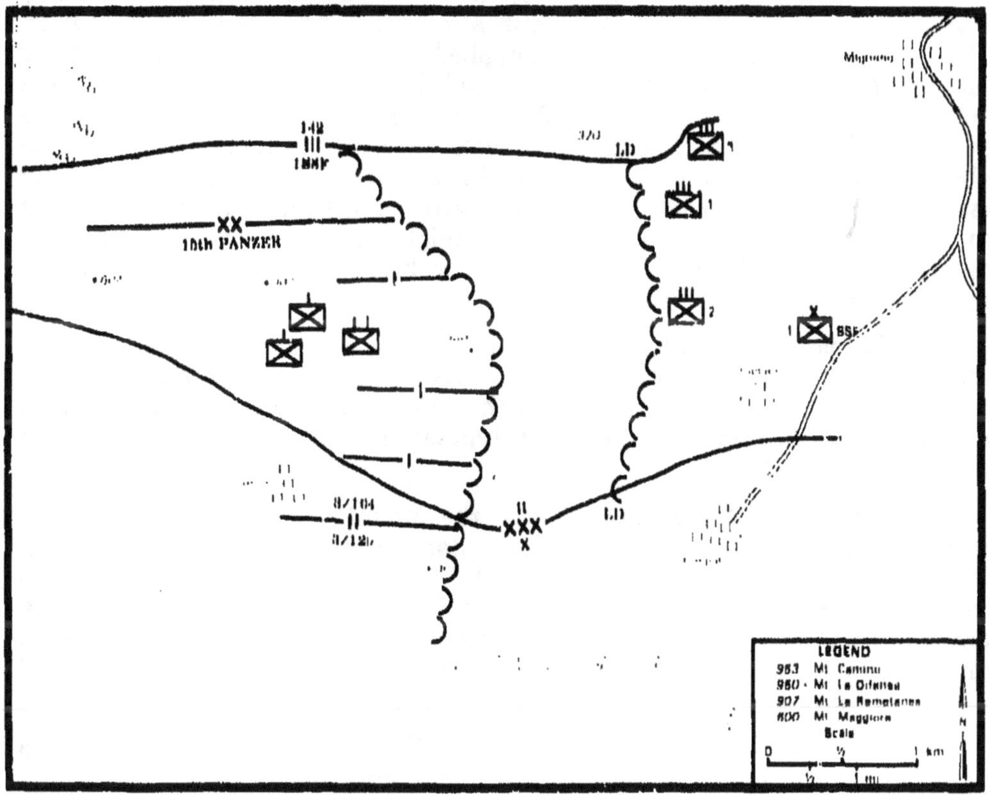

Source: C3I Battlebook 14-A, *Monte la Difensa*, U.S Army Command and General Staff College, Ft Leavenworth, KS, May 1984

Map 13 The Mount de la Difensa area

Allied forces were very steep, rough, jagged, and mostly bare of cover and concealment, except for ravines or gullies that traversed them and the scrub pines that dotted the lower elevations.

The only approaches into this inhospitable wall of rock and stone were primitive trails that were covered by German fires. Mount de la Difensa (960 meters high) formed the protruding corner of the hill mass, with Mount Camino (963 meters) to the south and Mount de la Remetanea (907 meters) to the west. Difensa's slope progressively ascended, finally terminating in perpendicular cliffs 200 feet high. Previous attempts to take the mountain had avoided these cliffs. Even the local villagers had deemed them impassable.

In November 1943, after ten or twelve days of rigorous, constant attacks on the enemy, the U.S. Fifth Army's advance ground to a halt against these formidable, barren bulwarks. The 7th Infantry, 3d Infantry Division, was thwarted in its assaults by the narrow approaches to the mountain, which severely limited the size of the point of attack. Even small toeholds on the mountain could not be held because of German snipers, machine gunners, and grenadiers, who dislodged every advance. The 7th Infantry was forced to withdraw with heavy casualties.[21]

The Germans had profited from strong fortifications, which were skillfully combined with natural obstacles. Their machine guns and mortars were dug several feet into the rock, making them almost invulnerable to artillery fires. Their positions, moreover, were well supplied, well camouflaged, and mutually supporting. Dispersed German snipers occupied hideouts from which they often held up unit movement for hours. The Germans also laid mines on all trails and natural approaches. Forward observers, moreover, called on the considerable amounts of German artillery available to fire accurately on anything seen moving below. In addition, German forces on one hill could easily support their neighbors so that an attacking unit might receive murderous fires from several directions. There was no easy way to break into this kind of fortified, interlocking defense.[22]

On Difensa proper, the Germans disposed the 3d Battalion, 104th *Panzergrenadier* Regiment, and half of the 3d Battalion, 120th *Panzergrenadier* Regiment (a total of about 400 men), with the 115th Reconnaissance Battalion in reserve. These veteran units had good reputations and could be depended on to defend staunchly.

The new Fifth Army plan to take the Mount Camino mass had three components (see map 14). General Mark Clark directed the X Corps to attack Camino in the south and directed the II Corps, with the FSSF attached, to capture Mount de la Difensa and Mount Maggiore in separate, but simultaneous, attacks. The II Corps commander ordered Frederick against Difensa and assigned Maggiore to the 36th Division. All attacks would be preceded by several days of heavy air and artillery bombardments.

The plan caught the Germans off guard. Not expecting the Fifth Army to try to storm the heights a second time, they anticipated a push by armor-heavy forces through the Mignano Gap, and they had positioned their reserves accordingly. Later, the Germans complimented the creators of the new plan for its cleverness.[23]

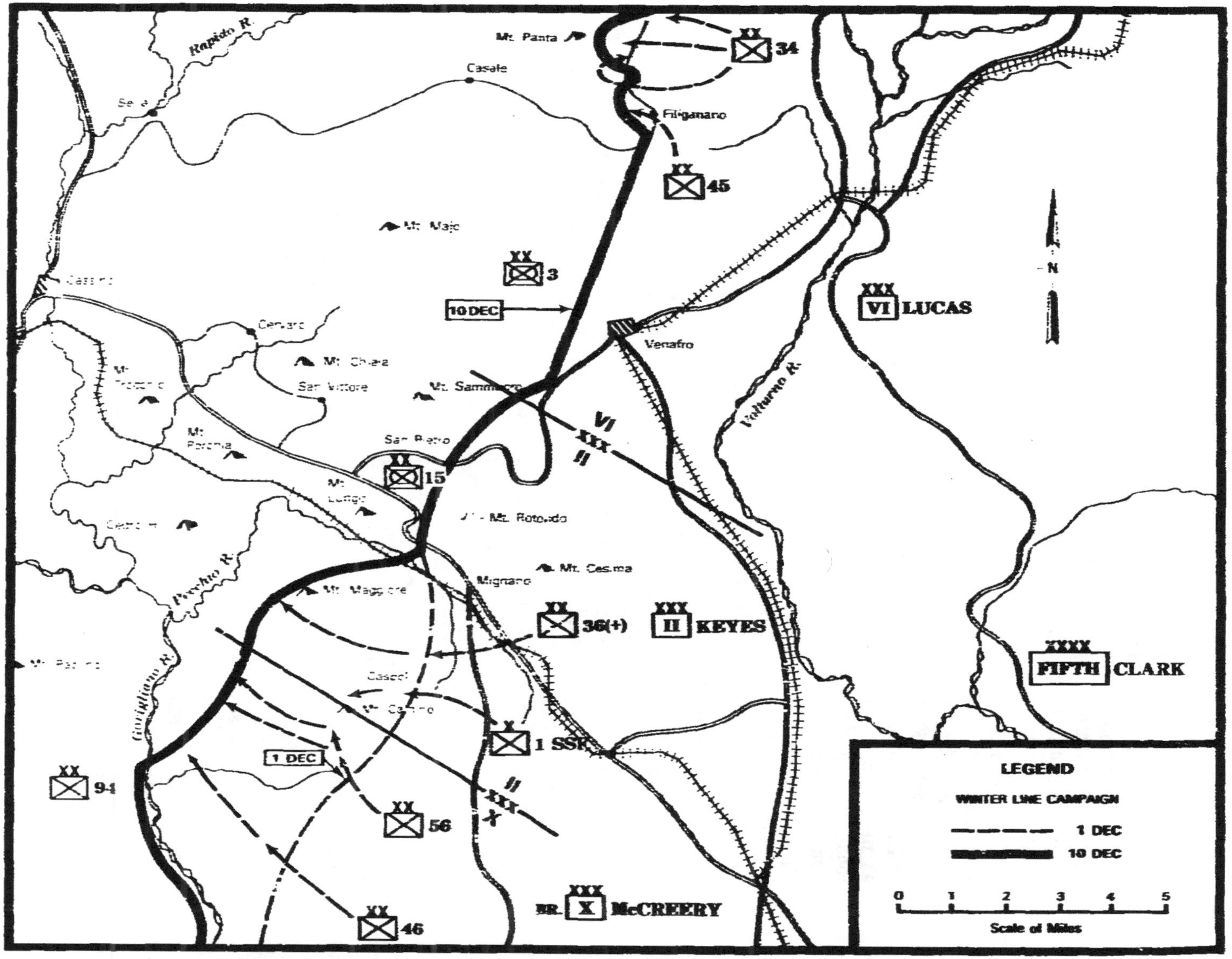

Source: CSI Battlebook 14-A, *Mans to Défense*, U.S. Army Command and General Staff College, Ft. Leavenworth, KS, May 1984.

Map 14. The clearing of Mount Camino, 1–10 December 1943

The FSSF was attached to the 36th Division from II Corps for the attack. The Force's mission was to attack and seize Difensa on the morning of 3 December and to continue the attack, on order, to seize Remetanea. Major General Walker, the 36th Division commander, retained the 1st Regiment as division reserve. Frederick assigned the assault mission to the 2d Regiment and ordered the 3d Regiment to position its 1st Battalion at the 600-foot level on the mountain to be the FSSF reserve. The 2d and 3d Battalions, 3d Regiment, were ordered to assist the Service Battalion in the resupply of the assault units.

Frederick also decided to attack the mountain at night from the northeast side, straight up its sheer face. He knew from the experiences of the 7th Infantry that the enemy had the other approaches from the north and southeast well covered by observation, fires, and mines. He felt confident in the abilities of his men to negotiate the difficult climb, and he believed that the approach at night, in a direction assumed impossible, would achieve surprise both in location and time. Surprise was essential to his plan. He also counted on heavy air and artillery bombardments to keep the Germans' heads down and focus their attention on the conventional approaches to the summit. This plan, if executed properly, promised the capture of the peak by shock, rather than by a long, drawn-out battle of several days' length.

On the evening of 1 December, the regiments moved out by truck in a light rain from their rearward assembly areas. Dismounting from the trucks, the men of the 2d Regiment trudged a hard ten miles through cold rain and mud to their preassault position about halfway up the mountain. Concealing themselves in ravines and scrub pines, the men waited during the next day, trying to stay warm and dry and to rest. Most officers were too busy to sleep, as they sorted out details concerning reconnaissance and supply. Simultaneously, the 1st and 3d Regiments moved into their designated holding areas (see map 15).

During the day and night of 2 December, Allied bombers and artillery delivered the heaviest concentration of indirect fires yet seen in the western war. Eight hundred twenty pieces of all calibers fired round after round of high explosive, white phosphorus, and smoke on the Camino mass. In a one-hour "serenade," 22,000 rounds from 346 pieces exploded atop Difensa.[24] While the preparation did not cause a great many casualties, it disrupted the German lines of supply to the front, destroyed wire communications, prevented the movement of reserves or shuffling of frontline troops, and limited the fires of German artillery.[25] These strong supporting fires continued from 2 until 10 December and severely hampered the efforts of the Germans to counterattack against the Allied ground forces.

At approximately 1800 on 2 December, the 2d Regiment began its ascent of Difensa for the assault, with 1st Battalion leading, the 2d Battalion in trail. As these units moved out, the 1st Battalion, 3d Regiment, also began its climb to its reserve position at the 600-meter level.

By 2230, the 2d Regiment reached the base of the Difensa crown. While the bulk of the regiment paused, scouts and rope teams continued up the final 100 to 200 meters to emplace rope ladders. The fog, wind, and freezing rain made for a bitterly cold night. Men huddled together in the darkness to

180

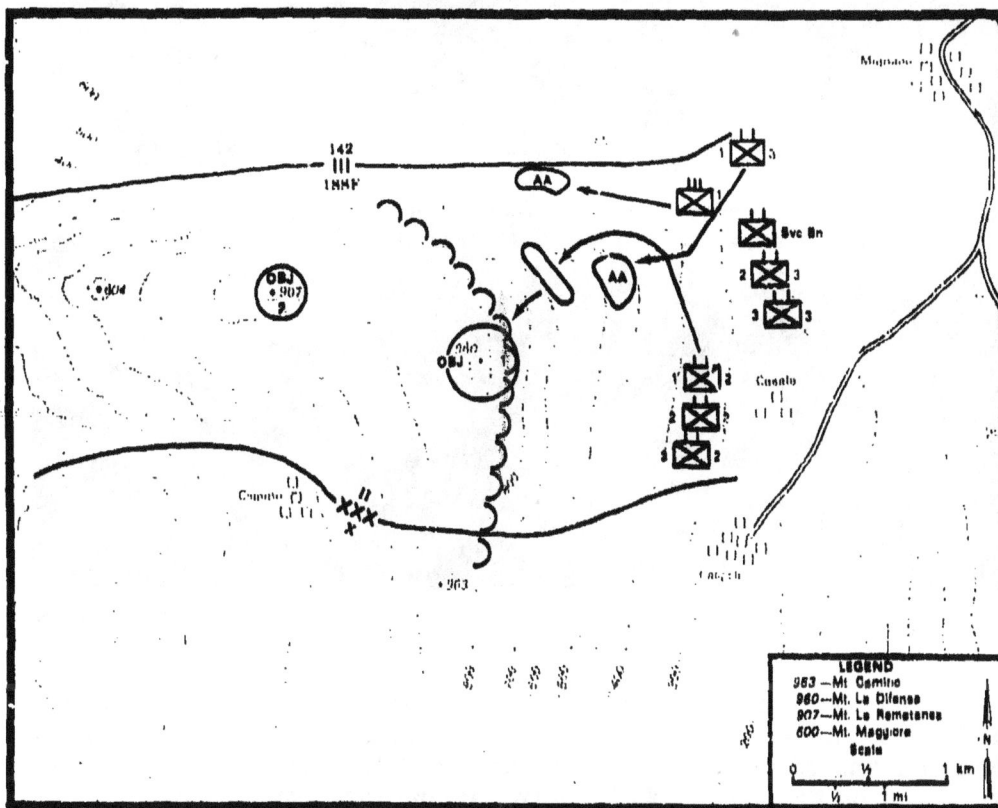

Source: CSI Battlebook 14-A, *Monte la Difensa*, U.S. Army Command and General Staff College, Ft. Leavenworth, KS, May 1984.

Map 15. FSSF movement to attack positions, 2 December

keep warm. Perhaps their main comfort was the sound of the increased intensity of the friendly artillery bombardment that pounded and illuminated the summit.

The units began to move again at 0100. The 1st, 2d, and 3d Companies, 1st Battalion, clawed their way up the rope ladders in two hours. From the top of the cliff, it was 350 more yards of steep, rocky slope to the actual peak. So far, the Germans had not detected the presence of the Force, even though several enemy artillery rounds landed sporadically farther down the mountain where the 2d Battalion was following. The difficulty of the climb is described vividly in *The Devil's Brigade*:

> The difficulty involved in this move comes into sharp focus when it is remembered the cliff face of Difensa begins at the 2,000 foot level and extends upwards at a pitch of 60 or 70 degrees for approximately another 1,000 feet. The peasants of the nearby villages never used anything but the well-cut trails on the opposite side when they took their flocks to the summit to graze. Since the winter of 1943, only two persons have tried to get to the top by means of the northeast face. These were two young men from Northern Italy, and the peasants who extricated them from the cleft in the rock wall where they had become stranded, cursed them for fools.
>
> This was the cleft that 600 riflemen, carrying packs which would have forced lesser men to the ground, negotiated without a sound. They groped for crevices with frozen hands while stretching their muscles to the aching point to keep from sliding backwards.[28]

181

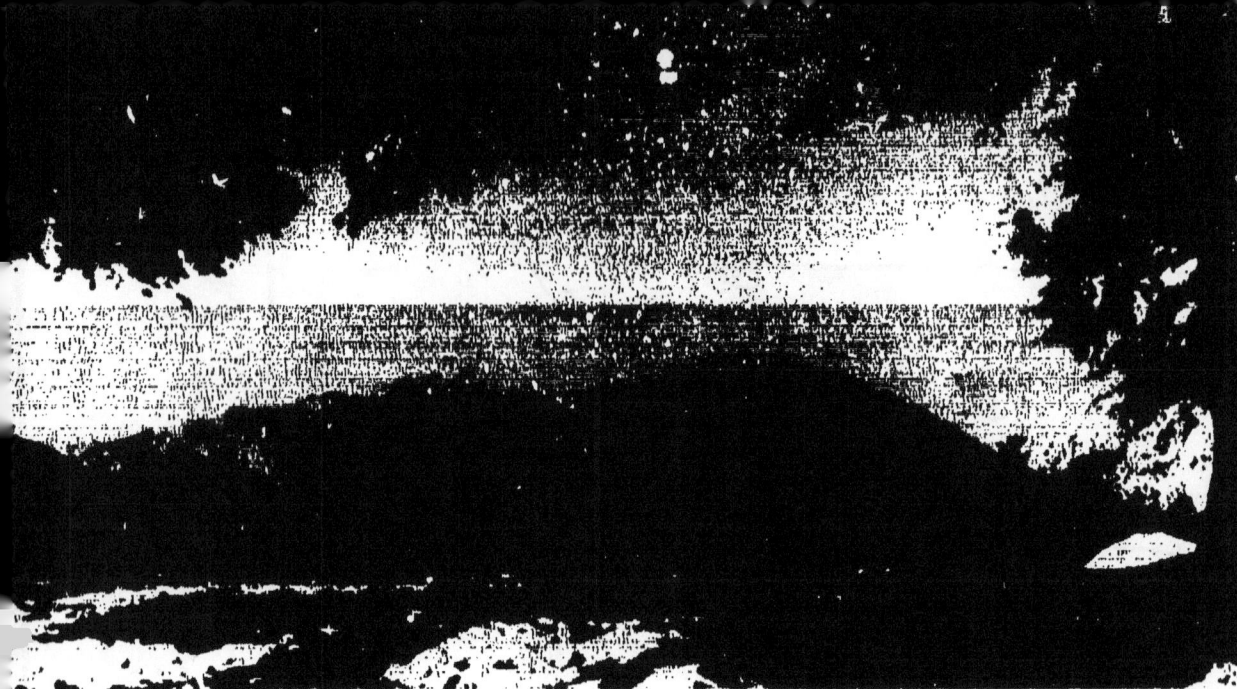

The Camino hill mass

One by one, the companies inched over the cliff and maneuvered into line for the final assault. Occupying the left side of the assault line was 1st Company, with 2d Company in the center. As 3d Company moved up to take its position on the right of 2d Company around 0430 (with 2d Battalion on the rope ladders), a rockfall alerted the Germans. Suddenly, the sky was full of flares, and German mortar bombs and machine-gun fire began to rake the men of the Force. At this point, the battle quickly deteriorated into a fight by platoon and section leaders. Yet sufficient surprise had been achieved to heighten the Force's chances for success. Moreover, the small-unit leaders in the Force had been fully briefed on the overall plan. As the battle fell into their hands, they knew exactly what to do. The Germans were hampered in their resistance by weapons oriented on the wrong lines of fire.

The 1st Company attacked first, supported by its own light mortars and light machine guns. As the sun came up, the company was well within the German defensive position with 2d and 3d Companies beginning to close with the enemy from their sectors. The men of the Force conducted fire and maneuver against each German strongpoint, suppressing it with fire, while other FSSF elements closed in on the enemy's flanks and rear. The unit leaders, including Colonel Frederick, were in the midst of the fight, leading, directing, and dying. As bits of fog blew away, elements of the Force found themselves suddenly exposed to snipers or enemy fighting positions. By 0700, the entire battalion was on the summit, and some Germans began to surrender, while their comrades streamed away to the west towards Remetanea. In under three hours, the 1st Battalion, 2d Regiment, had taken an objective that had kept the 3d Infantry Division at bay for ten days.

Low ammunition stocks prevented the regiment from continuing the attack to Remetanea. In addition, the men were exhausted. The 2d Battalion moved up to join the 1st Battalion, but Colonel Frederick held up any further advance until ammunition, water, rations, blankets, and litters could be packed up the mountain—a six-hour exercise at best. In the interim, the battalions on Difensa reorganized and consolidated their positions for an expected counterattack, prepared the wounded for evacuation, cleared out the remaining snipers, established outposts, and pushed a few patrols out to determine the disposition of the enemy defenses along the western ridgeline.

The next five days resembled an exercise in survival. As the trickle of resupply flowed in by packboard and the prisoners of war and the wounded moved downhill, the weather turned worse. Heavy rain fell steadily, day after day, with few breaks. Wet, cold, exhausted, and beginning to suffer from exposure, the Force confirmed Napoleon's maxim that "the first quality of the soldier is enduring fatigue and privations; valor is only the second."

To make matters worse, the British 56th Division, though attacking valiantly, had not taken the Camino peak and would not do so until the evening of 6 December. As a result, the FSSF had to endure intense mortar and long-range machine-gun fire from Camino as well as Remetanea.

At the bottom of the hill, Colonel Adams coordinated the steady but slow stream of supplies to the top. Mules could not handle the grade or the footing. Every can of water, every ration, every round of ammunition had to be wearily carried up by hand. The entire 3d Regiment, less one company, was dedicated to assist the Service Battalion in this effort.

Recognizing the limits these conditions imposed on his operation, Frederick decided to postpone his attack against Remetanea until 5 December. When Walker released the 1st Regiment, Frederick called up one of its battalions to Difensa to hold the summit so that the 2d Regiment could attack. On its way over, the 2d was delayed twenty-four hours and suffered 50 percent casualties in an attack by German artillery that might have been entirely accidental.[27]

From 3 to 5 December, the small forces on the summit continued to feel their way toward Remetanea. Because of the fog and the broken terrain, the fighting assumed no fixed pattern. A temporary break in the fog often found both German and U.S. forces helplessly exposed while they felt their way about the hill mass. At one point, 400 German reinforcements were discovered moving forward for a counterattack. Hastily arranged artillery fires foiled their attempt.

Finally, on 5 December, the 2d Regiment sent two reinforced battalions down the ridge toward Remetanea in a daylight attack (see map 16). They were stopped about halfway to the objective by bitter, desperate resistance. The German defense, however, dissolved during the night so that the regiment was able to occupy the peak against light opposition by noon on the 6th. Over the next two days, the FSSF cleared the area they held of isolated German snipers and outposts and tied in on the left with the British and on the right with the 142d Infantry. Relieved on 9 December, the Force wearily climbed back down the mountain to recuperate. It had suffered 511 casualties: 73 killed, 9 missing, 313 wounded or injured, and 116 incapacitated from exhaustion and exposure.[28]

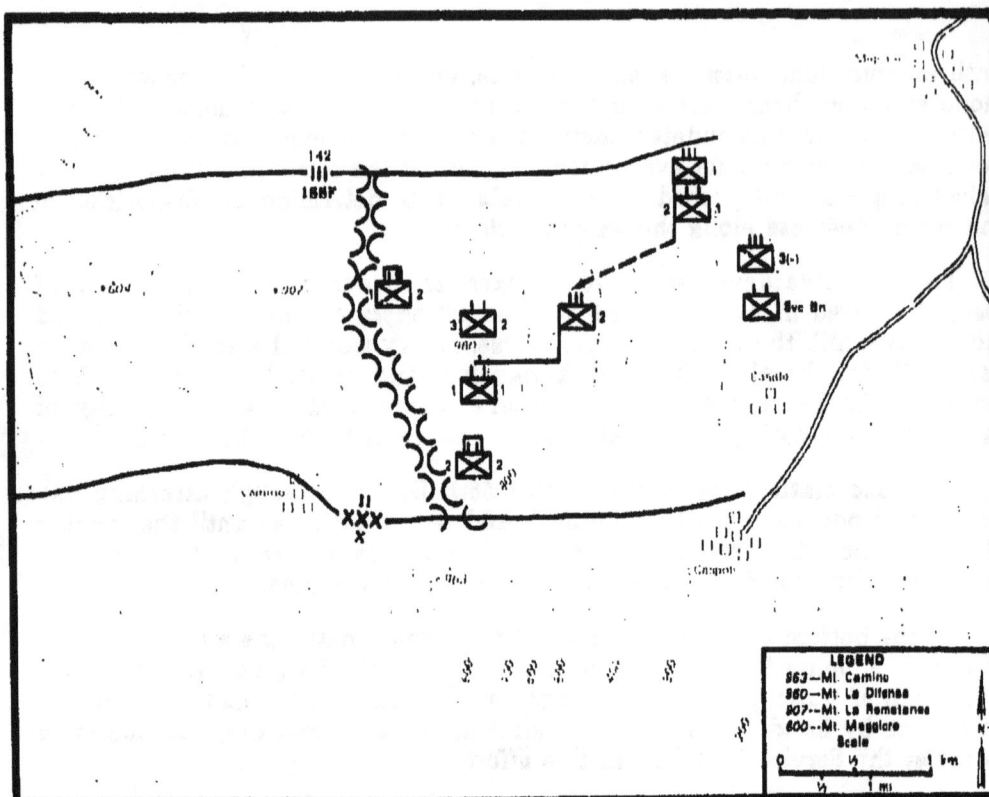

Source: CSI Battlebook 14-A, *Monte la Difensa*, U.S. Army Command and General Staff College, Ft. Leavenworth KS, May 1984.
Map 16. Clearing Mount de la Remetanea, 1943

Assessment of the Battle

The participation of the FSSF in the battle for the Camino high ground from November to December was crucial to the Allied success. Newly arrived in the theater in November, the over 100 percent strength of the Force enhanced its fighting power. The assaulting units of the 3d Infantry Division, at the approaches to Camino, had been exhausted and depleted from weeks of fighting. The special training, elite character, and high confidence of the Force gave it a further advantage over the 3d. What the FSSF accomplished in two to three hours at Camino is in sharp contrast to the failure of the 3d Infantry Division and is remarkable testimony to the capabilities of the Force. The Force's success in capturing Difensa's peak, moreover, contributed significantly to the conquest of the Camino and Maggiore peaks. (But even with this support, the 56th Division required five days to take Camino despite its use of excellent techniques of night attacks and multiple approaches.) Once Difensa fell, Camino, Maggiore, and Remetanea followed like dominos. Thus, the Force was the key used to unlock the entire bastion. Finally, the capture of the Camino mass, as the southern stopper in the Mignano Gap, led directly to the capture of Mount Sammucro, on the northern side of the mountain mass, in the last half of December.

The mission against Difensa was fully suited to the FSSF. It took advantage of the Force's special training in night fighting, mountain climbing,

Technician Fourth Grade Garbedian, radioman, on Difensa

cold weather, and lightning assault. No conventional unit, without special training, could have accomplished the mission.

Surprise and shock were the essential ingredients in the success of the Force. The risk taken in sending such a small assault element against the strong German positions was mitigated by the choice of the time and place of the attack. Furthermore, even though the Germans were not outnumbered by the attackers, they were overly dispersed and incorrectly oriented. The shock of the attack by the Force, achieved through speed and the volume of fire from its light weapons, overwhelmed the enemy and forced him to withdraw.

The Force attained surprise and shock only because of its specialized training and extraordinary fitness, which permitted it to climb the toughest part of the mountain; to mount a violent, rapid attack even though exhausted; and to endure debilitating extremes of weather and increasing exhaustion while continuing to hold, then expand, its position on the high ground. The superb fitness of the Force also allowed it to supply itself by man pack in the operation.

The Force's use of the cover of darkness and technical mountain-climbing skills permitted it to achieve surprise against a wary and experienced foe. Knowing the disposition and inclinations of the enemy, the FSSF turned the terrain to its advantage.

185

The leadership of the Force excelled in this, their first, live action. Frederick's plan was sound, its execution almost flawless. Once discovered, the junior leaders of the Force took charge of the battle and prevailed. Company commanders and their seniors led from the front. In this regard, a comment by a British officer who visited the FSSF at Difensa on 7 December is revealing. He remarked to a member of the Force his surprise at the number of majors and colonels present in the combat zone. The Force officer replied that both U.S. and Canadian members of the FSSF held the principle that leaders should lead.[29] On the other hand, Captain Pat O'Neill, the FSSF's expert in hand-to-hand combat suggested, "Perhaps we lost more officers than we should, as they needlessly exposed themselves."[30]

Frederick set an exacting example for the rest of his officers on Difensa. He was everywhere: carrying wounded men to the aid station or down the mountain, sharing a cigarette or a foxhole with his men, and going forward on personal reconnaissance. He was even seen praying beside a man wounded along the trail up the mountain. By the end of the war, Frederick would be wounded on nine different occasions, a record for general officers. It is also recorded that Frederick occasionally infiltrated an enemy position prior to an attack by his command. He was then able to observe from a covered position which of his subordinates were actively leading the attack.[31]

The victory at Difensa established the reputation of the Force as an extremely capable and hard-hitting raiding force for mountain operations. It demonstrated that it was a unique organization with unusual capabilities not matched by regular units. Ideally, the Army should have used the Force judiciously for special missions that required its distinctive capabilities. Unfortunately, Frederick's superiors do not appear to have properly evaluated the uniqueness of the Force. After two weeks of rest and recuperation, the FSSF was sent back into line, this time not as a raiding force but as a separate brigade, fighting along conventional divisions in a sustained role.

Mount Sammucro and Mount Majo

Following the battle for the Camino mass, the U.S. 36th Division shifted to the north and assaulted Mount Sammucro (1,205 meters), the northern shoulder of the Mignano Gap. After taking the peak with heavy casualties, the 36th Division still faced the prospect of clearing several miles of lower slopes on the western and southwestern sides of the peak. On 22 December, Major General Keyes, II Corps' commander, ordered the FSSF into the line against these positions. Keyes also directed the 141st Infantry and the 504th Parachute Infantry Battalion to support the FSSF, and he placed the 6th Armored Field Artillery Group and three airborne artillery battalions in direct support. The 36th Division Artillery was available in general support. On their part, the Germans defended from their usual well-fortified, mutually supporting positions, and they, too, held strong artillery forces in support.

Following a one-hour FSSF artillery preparation, Frederick planned to conduct a coordinated night assault against the Germans, but a German counterpreparation caught the assault battalion in its start positions, completely disorganizing it. As a result, the Force's attack did not begin until dawn. Because of the nature of the terrain and the enemy's dispositions, the attack took the

form of a frontal assault through enemy fire. For two days, from 24 to 25 December, the Force doggedly pressed forward, the 1st Regiment bearing the brunt of the fight. Again, the speed of its attack surprised and overwhelmed the Germans, but the cost to the Force was high; several companies were reduced to 20 to 30 percent strength. Nonetheless, the 1st Regiment took its objectives on 25 December and was relieved in place by other elements of the 36th Division. Except for the exceptional perseverance of the Force in the face of the stiff German resistance, nothing distinguished this battle from a hundred other bloody contests for Italian terrain.

After three short days of rest, the men of the Force hoisted their packs anew for their next operation. Having secured the Mignano Gap, General Keyes now intended to push his II Corps down the valley toward Cassino, with the 34th Division making the main attack. Because a vast area of high ground on the right threatened this advance, Keyes ordered the FSSF into the mountains to protect his flank. The Force retained the 456th Parachute Artillery Battalion in attachment and received an engineer company in support to clear mines and improve trails. A Sardinian pack mule company was also attached for logistic support. The 142d Infantry followed to occupy the ground cleared by the Force.

Essentially, this operation called for a wide end run through a sparsely settled and thinly held maze of barren hills and ridges. German opposition consisted of scattered outposts leading back to a main line of resistance occupied by about three battalions.

187

Frederick recognized that the large area required a broad advance during which his men would attack, occupy, and then be relieved from a succession of peaks and ridges. Accordingly, he organized the Force into two columns (see map 17). On the right, at about two-thirds strength, the 3d Regiment was ordered to sweep northward, then eastward toward Mount Majo (1,259 meters). On the left, Frederick directed the 1st and 2d Regiments (each at 50 percent strength) to work in tandem as they advanced parallel with the 3d Regiment toward Hill 1109. Frederick orchestrated the movements of the separate columns so that they would normally be able to support each other from over-watching positions. The 2d Regiment also detailed three of its companies to assist the Service Battalion.

By this time, winter had descended on the area in full fury. Ground above 600 meters was covered by five inches of snow or more. Temperatures remained below freezing, and high winds caused the chill factor to plummet.

Force scouts moved out on 1 January 1944. (Commanders typically preceded each movement with their own reconnaissance.) The 2d Regiment entered the defended area on the night of 3 January. The first objective, Hill 724, fell to the attackers after a short fight. The defenders had been completely surprised by the stealthy approach of the attackers.[32] On the evening of the 4th, the 1st Regiment duplicated the feat on Hill 675. Said one amazed German, "We were standing alertly at our machine guns when a voice said, 'Hands up!' The attack was very excellently accomplished."[33] Surprised by the speed of the Force's advance, the Germans in this zone fell back to Hill 1109.

If anything, the 3d Regiment moved even faster, also capitalizing on its stealth and night-fighting abilities to secure a string of nameless hills. On one hill, an FSSF company of 59 men crept to within hand-grenade range of a company of Germans about 100 men strong. Attacking at 0500, the outnumbered men of the Force annihilated the surprised Germans.[34]

The Mignano Gap

MT. CAMINO   HIGHWAY 6   MIGNANO   MT. SAMMUCRO

Patrolling by day and attacking by night, the FSSF rapidly closed on their final objectives. Lieutenant Adna Underhill explained the reason for the shock of the Force attacks:

> We like to attack because we don't get too many casualties then. We go straight into the Jerries until we get close enough to use grenades. We never let machine guns stop us or pin us down because we know that once we are stopped, we'll be murdered by their mortars. As long as we can keep going forward, there is less chance of getting hurt.[35]

The effects of the weather and German fire, however, were sapping the Force's strength. Keyes realized that the FSSF could not continue to advance on such a broad front without some help. Thus, on 6 January, Frederick received two battalions of the 133d Infantry and another engineer company in attachment. The next day, the entire 36th Division Artillery was placed in direct support. Combined with the FSSF, these reinforcements formed Task Force B, under Frederick's command.

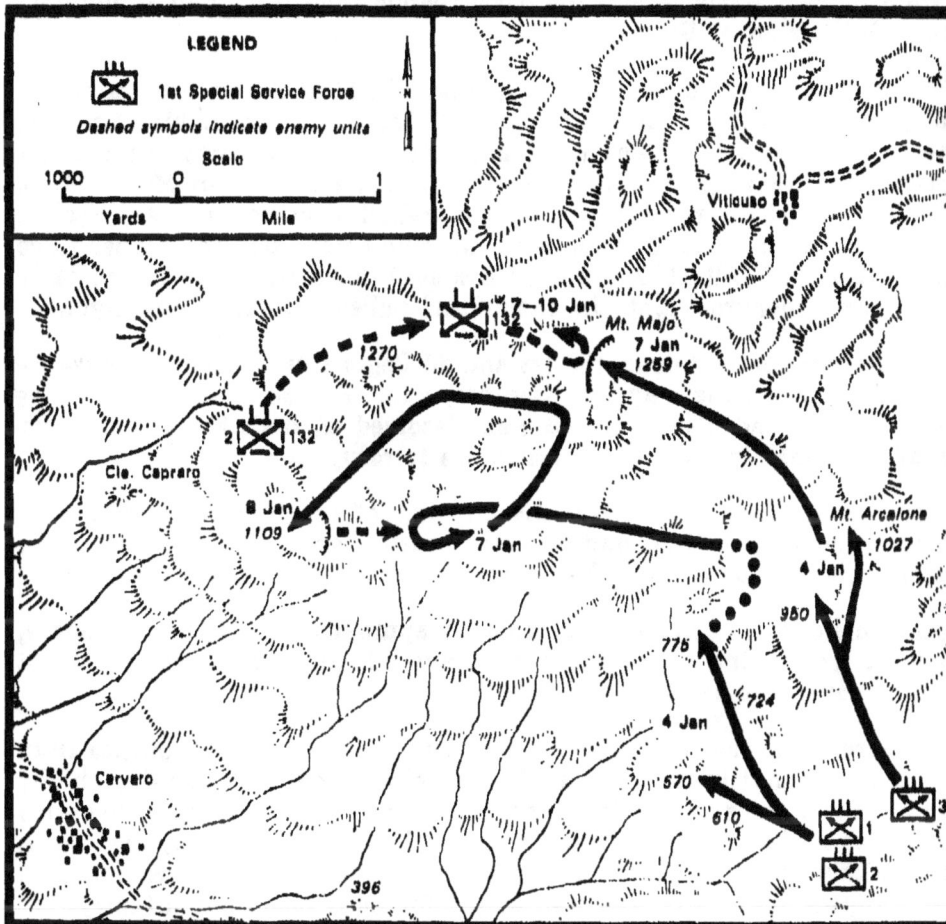

Source: U S War Department, General Staff, *Fifth Army at the Winter Line (15 November 1943—15 January 1944)*, American Forces in Action Series (Washington, DC: Military Intelligence Division, U S War Department, June 1945)

Map 17. The capture of Mount Majo and Hill 1109, showing the II Corps' right flank, 4—10 January 1944

Colonel Frederick decided to conduct separate but simultaneous attacks on the night of 6—7 January to secure both Mount Majo and Hill 1109 (see map 17). Colonel Walker, 3d Regiment commander, organized a two-pronged attack against Mount Majo. One battalion drove directly for the peak, while the other circled to the far side and attacked from the west. Approaching with their usual stealth, these elements executed the coordinated fire and maneuver so characteristic of the Force. They quickly closed to within tens of meters of the flanks of the German positions and pinched them off one by one. By 0520, Walker's men had occupied the summit; by 0900, they took the neighboring high ground.[36] This feat was especially remarkable in that the force ratio between attackers and defenders was approximately 1:1.

For the next three days, the 3d Regiment held Mount Majo against twenty-seven separate counterattacks. In this defense, the FSSF made good use of a number of German machine guns that had been left behind with a large stock of ammunition. The better part of a German regiment was crippled by fires from its own weapons and the 8,500 rounds fired from 7 to 10 January by the 36th Division Artillery.[37]

The 1st Regiment had less success against Hill 1109. Strong counterattacks forced it off the slopes on the morning of 7 January. Colonel Marshall, the regimental commander, decided to plan an easier attack, using the indirect approach. That night, he moved his regiment to the vicinity of Mount Majo, now held by the 3d Regiment. From there, Marshall attacked westward against Hill 1270, seizing it by 0200, 8 January. Marshall now held the higher ridge to the northeast of Hill 1109. His men attacked down the ridge in the early morning. Resistance was light; the enemy had pulled out during the night.

Over the next several days, the 1st and 3d Regiments assisted the advance of Task Force B all the way into Cervaro, with the 133d Infantry carrying the brunt of the load. Task Force B was dissolved on the 13th, and the remnants of the FSSF limped back to rest areas to recuperate.

Like the conquest of Difensa, the clearing of the hills between Mount Sammucro and Mount Majo had a major bearing on the outcome of the II Corps offensive. In particular, the occupation of Majo collapsed that part of

the German line guarding the southeastern approach to Cassino. Certainly, the main forces in the valley could not have advanced so rapidly had not their right flank been secure.

The Force's mission at Majo was more in line with its capabilities than with a conventional infantry unit. The mission capitalized, again, on the Force's mountain skills and on its experience in cold weather. But the operation had more of a meat-grinder character to it than the earlier experiences at Difensa (although not as much so as the operation on Mount Sammucro). The Majo operation clearly was well suited for light mountain infantry, but one must remember that the FSSF had training and skills beyond those of normal light infantry. Each casualty experienced by the Force represented not only the loss of a cold-weather mountain expert, but it also represented the loss of a skier, paratrooper, demolitions expert, and amphibious raider. So

190

even though the FSSF performed its tasks in these Italian hills better than any conventional unit could have hoped to, the operation remains to some degree a waste of the special capabilities of the Force.

The Force's casualties were very high. On 8 January, combat strength in the Force stood at 53 officers and 450 enlisted men, with the 1st Regiment down to 7 officers and 82 men. By 14 January, the Force had experienced 1,400 casualties out of the 1,800 men in its combat echelon. Morever, Service Battalion strength had decreased to 50 percent from fatigue, wounds, and deaths. Many of these casualties resulted from the extreme weather. The Canadian war diary tells the grim story:

> SANTA MARIA: 2 January. A bright cold day. Parkas are being sent forward as there is about 5 inches of snow in the hills and quite cold.
>
> 7 January. Bright and cool. Casualty returns from the front include a number of frost-bitten feet.
>
> 8 January. Continues bright and cold. Today's casualty return from the R.A.P. lists nearly 100 names, half of them frost bite and exposure, the rest battle casualties. The weather in the hills is very cold, high wind and snow. German resistance is quite severe, artillery and mortar fire still taking its toll.
>
> 9 January. Today's Force casualty return has 122 names. Again nearly half are frost bits and exposure. There won't be much left of the Force if casualties keep at this rate.
>
> 10 January. Mild and damp. News from the Front is bad. The Force is being thrown into one action after another with only a handful of able-bodied men left and no sign of their being relieved; 73 names on today's casualty report, 40 frost-bitten feet. Those returning to camp on light duty say it is really rugged and they are all played out. Three weeks tomorrow since they left here.[38]

The number of casualties attributed to frostbite seems unusually high. It may well be that the élan and bravado of the Force kept it from taking reasonable precautions against injuries of this nature. Frostbite can almost always be prevented through proper care, buddy checks, and supervision. An after-action report from the Force admits to a certain laxness in this regard. This report recommended the "constant daily care of body and feet whenever and wherever the opportunity offers." It went on to describe a situation where two companies were fighting on the same high ground in bad weather: "One company commander made it a personal task to see that every man removed his boots and rubbed his feet at least once per day no matter how intense the action; another company commander did not take the same precaution. The cases of 'trench foot' in the former were negligible; in the latter company high casualties from this cause resulted."[39] Obviously, close supervision by first-line leaders and junior officers is a must in the prevention of such injuries.

During the combat from Mount Sammucro to Mount Majo, the FSSF exhibited many of the same tactical techniques as they had at Difensa. Shock and surprise based on speed, stealth, silence, and violence (when contact was finally made) stand as the hallmark of the Force. Their attacks were almost exclusively at night, usually without preparatory artillery fires. Despite the disadvantage of unfavorable force ratios—less than 3:1, sometimes less than 1:1—the Force overwhelmed the enemy through shock. In attacks, it relied primarily on its individual and crew-served infantry weapons and hand grenades. When defending against German counterattacks, it wisely let the artillery do most of the work.

Because of the large area to be covered and the nature of the terrain, the Force was involved frequently in decentralized company operations against separate objectives. Frederick, to maintain his forward momentum and secure his rear in such operations, often needed combat support from the engineers and artillery and reinforcement from other infantry. However, no tank or anti-tank support was yet required, since he had come up against no enemy armor in the very restrictive terrain.

In their operations, the FSSF captured prisoners almost at will for the specific purpose of immediate tactical intelligence. This technique is a testimony to its confidence, daring, and superior field craft. The Force also exhibited endurance, fortitude, and esprit, which kept the unit moving forward even after its casualty figures exceeded the normal ceiling for combat effectiveness.

The Anzio Beachhead

On 30 January 1944, General Clark ordered the FSSF into the Anzio beachhead, where it quickly moved into the right-hand sector of the defensive perimeter. The right-hand portion of the beachhead was the quietest sector. The Germans maintained only a holding force there, while they violently attacked the center and west side of the beachhead. Thus, in one sense, the defense of the eastern part of the perimeter can be considered an economy-of-force opera-

Mount Sammucro, with San Pietro on the right

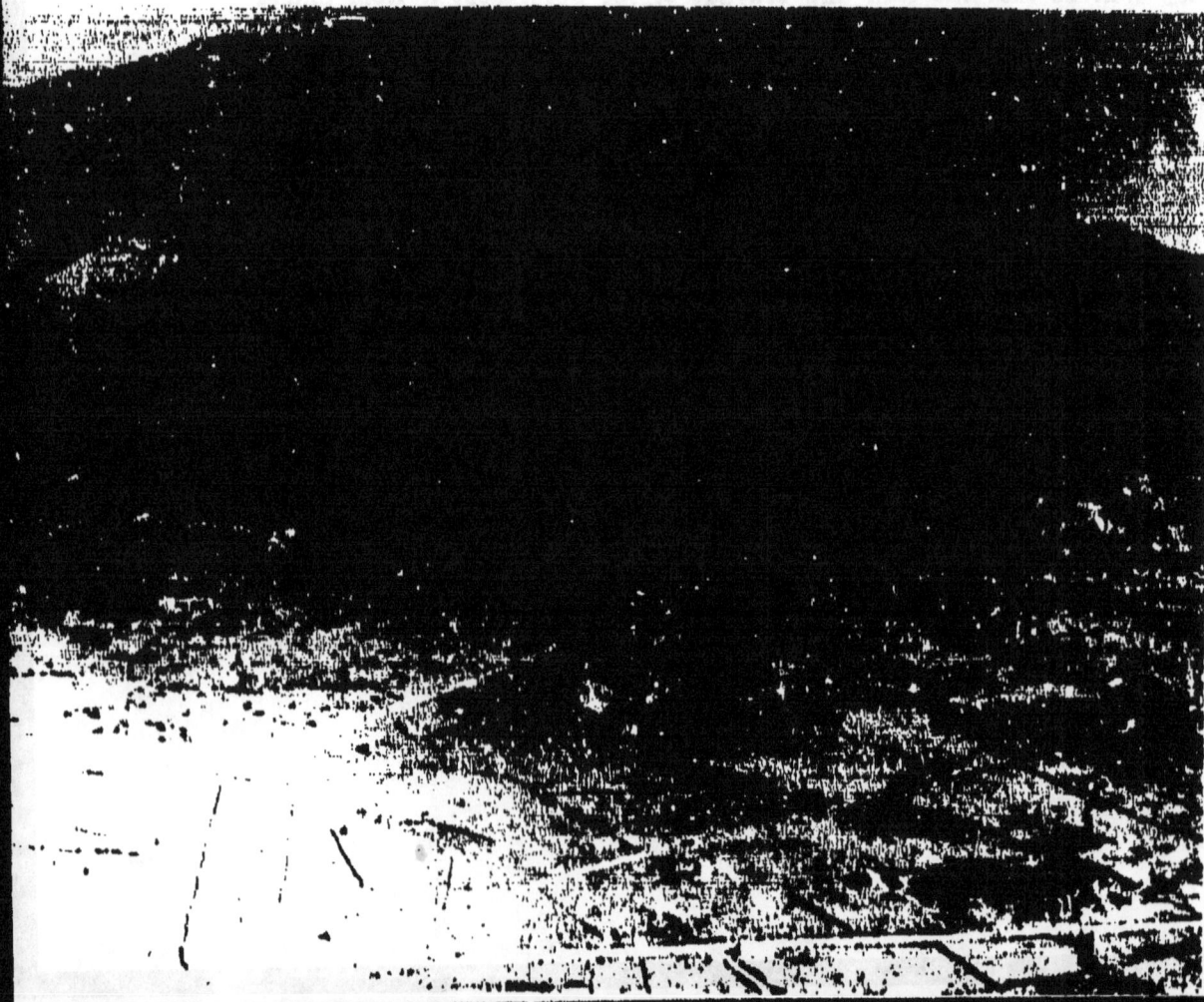

tion appropriate to a light infantry force properly dug in and augmented with artillery. Still it was a surprising mission for an organization like the Force, an organization designed and trained for offensive raiding. Defending the Anzio beachhead put to use only a fraction of the Force's capability. (The dispositions of the units in the beachhead are shown in map 18.)

Promoted now to brigadier general, Frederick assigned the 3d Regiment to the upper eight kilometers of the line that ran along the Mussolini Canal. The 1st Regiment occupied positions along the canal from the 3d Regiment's right flank to the sea five kilometers away. The 2d Regiment formed the reserve. The 456th Parachute Artillery Battalion had moved to Anzio with the Force in an association that had now become all but permanent. The FSSF also received the 81st Reconnaissance Battalion (light tanks and armored cars) and some engineer elements in attachment. An additional two to three other artillery battalions supported the FSSF with their fires. The Force occupied this sector with little change for ninety-eight consecutive days.

The enemy forces facing the FSSF came primarily from the Hermann Göring Division. They numbered about 1,250 men organized into 2 provisional battalions plus separate companies, including a tank company of 12 to 15 Mark IV medium tanks. The enemy was also supported by artillery and 88-mm guns displaced forward into outposts.[40]

Despite its formidable opposition, the Force began to impress its unique personality upon the beachhead from the first night of its arrival. Taking over the line from the 39th Combat Engineer Regiment, the men of the Force discovered that the Germans had outposts right on its doorsteps, an intolerable condition. That night, the 3d Regiment sent out five patrols to start clearing a buffer zone between the lines. Crossing the canal on improvised foot bridges, the patrols infiltrated to a depth of 1,000 yards. The next day, the artillery used the information the patrols obtained on enemy locations to bombard the Germans. The 1st Regiment followed suit with similar patrols the next night.

In this fashion, the Force initiated a pattern that it repeated over the next several weeks. Every night, strong combat patrols eliminated the forward enemy outposts, using the techniques of surprise and attack that they had perfected in the mountains. The Germans did not like it. They soon pulled their outpost line back a distance of 1,500 to 2,000 yards, granting the Force the breathing room it desired. By seizing the initiative in this manner, the Force immediately established its moral ascendancy and tactical superiority over the Germans.

Once the respective defensive lines stabilized, the fight in the eastern sector became a contest to see which side would control the no-man's-land that now existed between them. The terrain had an unavoidable influence on this struggle. The area occupied by the Force and immediately to its front was billiard-table flat. The high water table prevented the digging of foxholes except on the high berm of the canal itself. Protection was available here and there in the form of infrequent tree plantings, scattered farmsteads, and drainage ditches. Farther back, the Germans occupied the hills and low mountains rising above the beach. Neither side could move about on the flat terrain during the daytime without attracting artillery fire. At night, however, the snakes came

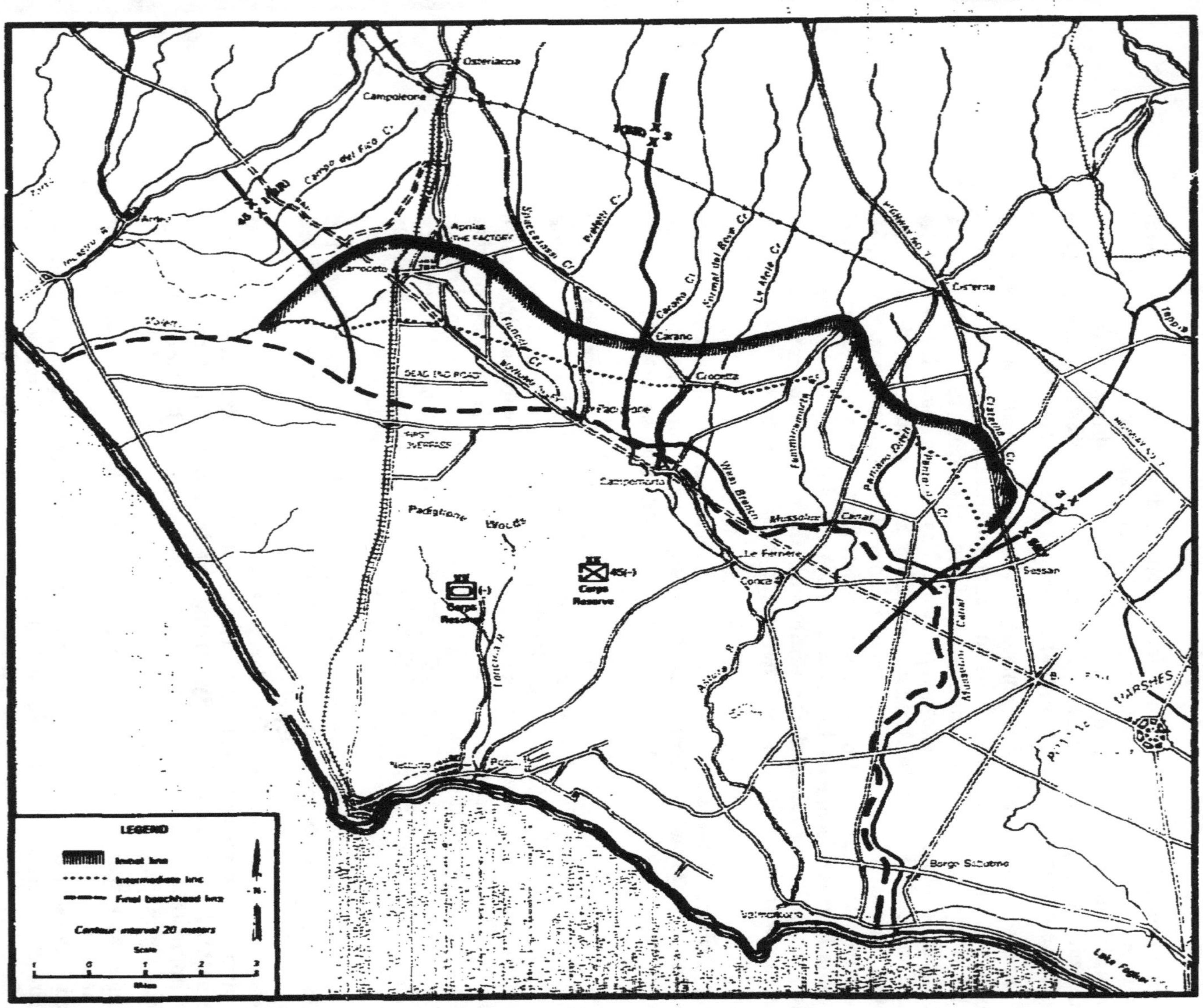

Source: U.S. War Department, General Staff, *Anzio Beachhead (22 January–25 May 1944). American Forces in Action Series*, Washington, DC, 1947, map 7.

Map 18. The plan for the beachhead defense, 5 February 1944

A 2d Regiment bazooka in action, Cerreto Alto

out of their nests. Force patrols moved out to occupy nighttime outposts and to explore and drive the Germans out of theirs. The Germans tried to do the same. The result was a fluid situation in which small bodies of men laid traps and bumped into each other as they circulated within the disputed ground. The men of the Force truly enjoyed this game of cat and mouse—and they were very good at it.

The night combat also took the form of raids and deliberate attacks by the enemy. But German attacks on the canal defensive line were rare and ineffective; the Force detected them early and handled them easily.

Force raids were designed primarily to keep the Germans off-balance and to demonstrate how vulnerable they were. Initially, the raids were small in scale, with specific objectives. For example, on 10 February, a composite company from the 2d Regiment seized the town of Sessuno and held it throughout the night. A week later, another company raid against a suspected enemy ammunition dump near Sessuno resulted in fourteen enemy dead.[41] In late February, a reinforced platoon set a trap in a row of houses used by the Germans to shelter patrols. Over the course of the night, the platoon captured several groups of enemy soldiers, totaling 111 men, at the cost of 2 men wounded. These prisoners told their interrogators that the Germans believed that the Force was a division, apparently because of the strength of the defense and the ferocity and frequency of the night patrols.[42] The raids were not all one way. The Germans occasionally chased the men of the Force out of buildings that they had occupied as outposts. Thus, the nights were busy with activity; the artillery ruled by day.

In April, however, Frederick decided to escalate the action through the mounting of a number of larger-scale daylight attacks against specific, limited

objectives. During the second week of April 1944, the 2d Battalion, 2d Regiment, combined with a platoon of medium tanks, a platoon of tank destroyers, and an assortment of light tanks and armored cars from the 81st Reconnaissance Battalion to form three strong armor-infantry company teams. Supported by artillery and mortars, each company team moved across the canal at night and attacked three separate objectives near the coastline at dawn. Altogether, the raiders captured sixty-one prisoners, killed an estimated nineteen soldiers and burned houses the enemy had been using as shelters. The raiders penetrated to a depth of about four miles and withdrew about 0900.

On 18 April, Frederick directed another raid, this time in the north. A Force company reinforced with a platoon of tanks crossed the canal in the early morning and moved into ready positions. At dawn, the company attacked a string of houses organized into a strongpoint by the Germans. The company overwhelmed the strongpoint, destroyed a number of enemy automatic weapons, took eight prisoners, and withdrew, having suffered only one minor casualty.

These raids had a number of benefits. They demonstrated the tactical superiority of the Force and instilled a defensive, wary mind-set in the enemy. They also inspired other units in the beachhead by fostering a spirit of aggressiveness and offensiveness. Some VI Corps units contacted the Force and asked its officers to lecture their men on raids and patrolling. In addition, news reports from the beachhead emphasized that not only were the Allies holding, they were also giving the enemy a bloody nose. Furthermore, the intelligence acquired during the raids enabled the artillery to attack enemy dispositions more accurately and enhanced the planning for an eventual breakout. Finally, the raids provided the Force with valuable experience in the conduct of tank-infantry operations, experience that would prove useful to it in the breakout from Anzio and the advance to Rome.

The men of the Force were also able to use their demolitions expertise during their operations at Anzio. Blowing up houses became a characteristic patrol function. The targets were houses in no-man's-land that were used at night by the Germans as strongpoints or observation points. The FSSF also blew up culverts and created road craters to impede any attempt by the enemy to conduct a strong armor attack against the canal.

Another interesting aspect to Force operations during this time was its use of local resources to increase the quality of its life along the canal, particularly regarding diet. Members of night patrols often brought back chickens, eggs, rabbits, goats, and other livestock for personal use. Others bought livestock and seed from local farmers. Crops were actually planted and tended; Force soldiers laid down their weapons for a time and took up the plow. Service Battalion even had its own chicken-cleaning plant. The 2d Regiment managed a small herd of milk cows. This kind of improvisation occasionally reached comical proportions, such as when two prisoners were marched in carrying a mattress for some lucky soldier and pushing a wheelbarrow full of potatoes and chicken crates. Through these productive efforts, the FSSF used the long daylight hours in improving the quality and variety of its rations.[43]

During its three months on the defensive perimeter at Anzio, the FSSF lost only 54 killed in action, 51 missing in action, and 279 wounded in action. Simultaneously, it received 53 officers and 1,408 enlisted men in replacements. These replacements fell into four categories: former Forcemen, now recovered from wounds and injuries; specially trained Canadian replacements from training centers in North America; the remnants of two Ranger battalions that had been disbanded after terrible losses in a failed deep raid at Cisterna; and carefully selected Americans from replacement depots. Because of the high quality and previous training of most of these men, they needed little additional training to come up to the quality of the Force.[44]

In retrospect, one could view the Army's use of the FSSF at Anzio as an opportunity to allow it to recuperate and renew its strength under relatively risk-free conditions while it performed the important service of securing part

Living off the land at Anzio

Courtesy of Colonel Robert D. Burhans

of the beachhead. The nature of the action also permitted the FSSF to sharpen its skills in patrolling and night fighting. On the other hand, once the Force had been committed to the beachhead, it was impossible to pull it out rapidly without endangering the lodgment. Thus, it was not available for any special operation more suited to its talents, had one arisen. A more imaginative Army commander might have found better use for the Force than the essentially static defense of the beachhead perimeter.

The Breakout from Anzio and the Drive to Rome

From January to May 1944, the forces in the Anzio beachhead had thrown back every attempt by the Germans to push them into the sea. Now, the balance of combat power had turned. New Allied divisions had come ashore, while the Germans had thinned their lines at Anzio to shore up other sectors farther east. Orders to prepare for a breakout of the beachhead went forward to all units, including the FSSF. On 9 May, the 36th Combat Engineer Regiment replaced the FSSF in the line. The FSSF pulled back into an assembly area and entered into an intensive retraining period. In particular, the Force needed to sharpen its edge of physical fitness (through exercises and marches), to perfect its assault tactics against strong positions, and to gain more experience in tank-infantry operations.[45]

The breakout plan directed an attack by three spearheads to pierce the German lines. The 3d Infantry Division made the main attack in the center to Cisterna. The 1st Armored Division on the left and the FSSF on the right were to advance parallel and cut Highway 7 on either side of the town. Thereafter, the 36th Infantry Division would pass through the 3d Infantry Division and take Cori, while the FSSF drove on Mount Arrestina to secure the right flank. The longer term objective of the VI Corps was Highway 6 in the region of Valmontane. From here, it was envisioned that the VI Corps would turn northwestward toward Rome (see map 19).

Reinforced by approximately two companies of tanks and two companies of tank destroyers, the Force initiated its attack at 0630 on 23 May, following a 1 1/2-hour artillery barrage. Around 1000, the leading regiment cut Highway 7, its initial objective. In advancing so rapidly, the Force had taken some significant casualties and had exposed its left flank, because the 3d Infantry Division had not been able to keep pace. The men of the Force had also advanced more rapidly than their supporting armor. Consequently, a German counterattack of infantry and twelve Mark VI tanks forced the FSSF to fall back a small distance. Faulty coordination (or execution) with the attached tank and tank destroyer units caused part of the problem.

At 0300 on 24 May, the 133d Infantry relieved the FSSF in place. That night, the Force continued its advance toward Mount Arrestino. By the evening of the 25th, the FSSF occupied the mountain, and the 3d Infantry Division pushed into Cori. The corps breakout had succeeded.

For the next two weeks, the VI Corps continued its advance to Valmontane and Rome. The FSSF was employed almost continuously in the role of a separate brigade on the corps' right flank, on high ground, to cover the main attack by a larger, heavier unit moving along the road nets. This covering

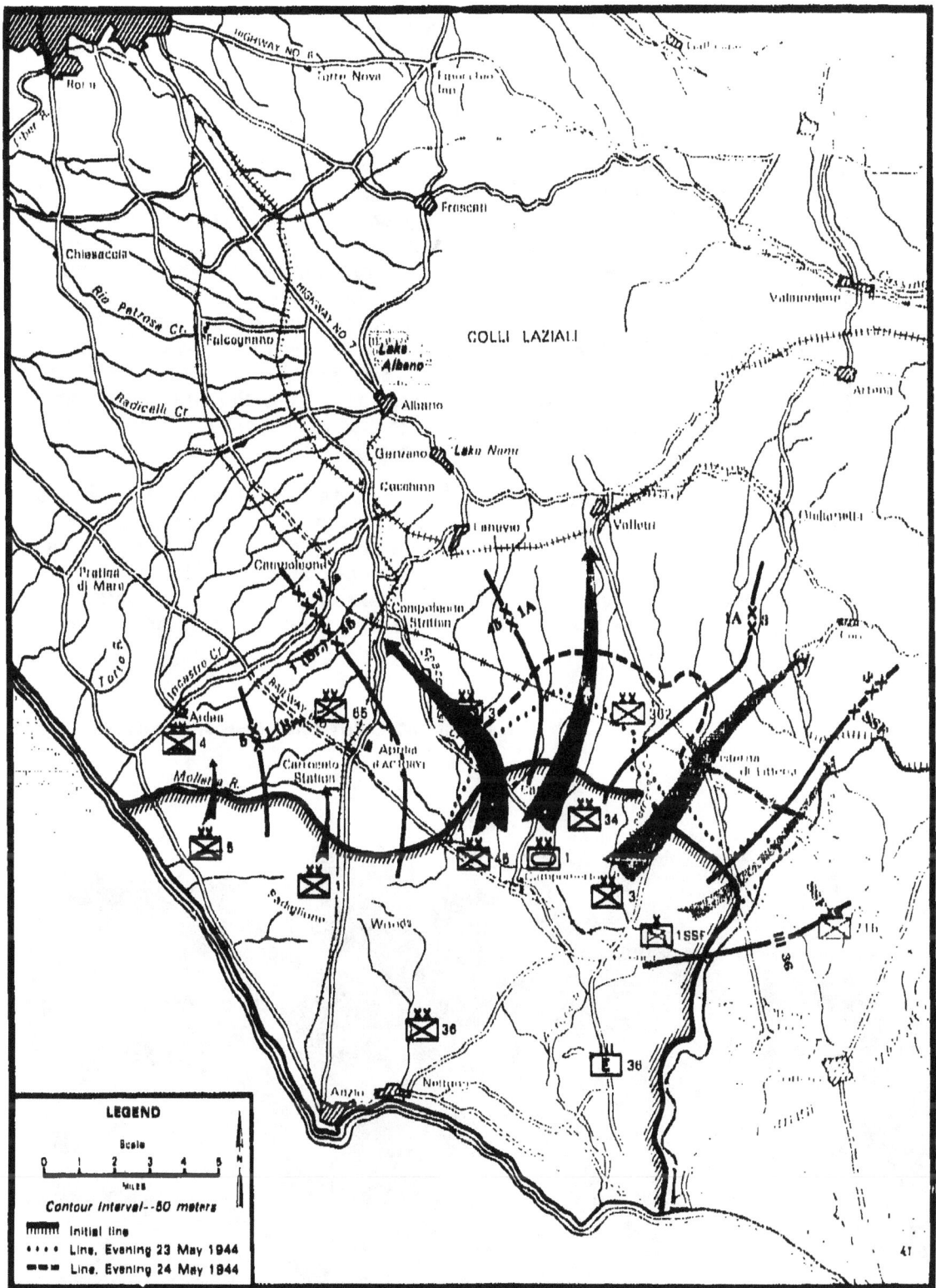

Source: U.S. War Department, General Staff, *Anzio Beachhead (22 January—25 May 1944)*, American Forces in Action Series, Washington, DC, 1947, map 8.

Map 19. The breakthrough, 23—24 May 1944

role usually meant that the Force advanced on a wider frontage. It also covered more distance, since it had to swing wider as the corps made its slow westward turn toward Rome.

These Force operations were indistinguishable from those of regular infantry brigades or divisions. Enjoying more combat support than it was accustomed to, the Force not only had its habitually attached artillery battalion but also acquired the Ranger Cannon Company (75-mm howitzers mounted on half-tracks). Adjacent division artillery and corps artillery also supported the Force with fires. In the breakout, the Force took on a combined arms structure, with tank destroyer, tank, armored-reconnaissance, and (sometimes) other infantry units. In order to give the FSSF more firepower, protection, and mobility, its task organization changed frequently, depending on the factors of mission, enemy threat, troops available, and terrain. Frederick and his regimental commanders showed exceptional skill in handling these task forces. In fact, it was probably Frederick's performance as a combined arms commander during the advance from Anzio to Rome that earned him his next promotion to major general and identified him as a likely division commander.

For the final dash into Rome, the corps attached Task Force Howze, a two-battalion, armor-heavy task force commanded by Colonel Hamilton Howze, to the Force to form a spearhead for the corps advance. The corps' order directed Task Force Howze to lead the advance by day and the Force by night. Frederick, however, later said that these orders were silly. Instead, as the senior commander, he used the armor and infantry together in a coordinated, continuous advance.[46]

Men of the 504th Parachute Infantry at the Mussolini Canal

Entering Rome proper, the Force had orders to secure a number of bridges over the Tiber River on the west side of the city. The situation in Rome was confused. The Germans had declared that Rome was an open city, yet they were defending it in different places in order to permit units retreating through Rome to escape from the advancing Fifth Army. Hidden strongpoints guarded many of the major intersections and the Tiber bridges. In addition, the Roman citizens were filtering out of their homes in anticipation of a generous welcome to the Allies, not realizing that they were clogging the streets and interfering with a rapid occupation.

Under these conditions, the Force—still coupled with Task Force Howze—moved through the city along multiple routes in small armor-infantry teams. These small elements avoided contact. Well-briefed officers and NCOs scouted out unguarded routes along which they might quickly lead their columns to the Tiber bridges. They cleared resistance where they had to and posted signs for the main body to follow. In this decentralized fashion, the Force (primarily the 3d Regiment) slipped past the German defenses and seized and held eight of the sixteen Tiber bridges. This feat of arms permitted the Fifth Army to continue pursuit of the Germans west and north of Rome and is a good example of the flexibility and initiative of the Force.

The Assault on the Hyères Islands

The FSSF held the Tiber bridges for two days, then turned them over to troops from the 3d Infantry Division on the night of 6—7 June. From Rome, the Force moved to a bivouac area beside Lake Albano for a well-deserved three-week rest. On 29 June, fresh orders alerted the Force to prepare for a new mission. The Seventh Army was planning Operation Anvil (later known as Operation Dragoon), the invasion of Southern France. The FSSF had been selected to be the spearhead of the invasion by conducting a nighttime amphibious assault against two islands flanking the invasion beaches. So, the Force moved to Santa Maria di Castellabate, a small fishing village in southern Italy to get ready for the operation. Colonel Edwin Walker was now the Force commander. General Frederick had earned a second star and been given command of the 1st Airborne Task Force.

Naturally, the nature of the impending operation influenced the training program of the Force. The naval convoys for the invasion planned to land three divisions of the VI Corps on beaches between Toulon and Cannes in the Cap Saint-Tropez area. In so doing, the convoys would pass to the east of two islands, Port Cros and Levant, which lay about seven miles off the coast. These islands had been occupied by the Germans in 1943. Aerial photography suggested the existence of a coastal battery with antiaircraft weapons on the western end of Port Cros. Levant, it appeared, had significant fortifications on its northeastern tip: three or four 164-mm guns, machine guns, pillboxes, a searchlight, and four medium coastal guns on the west end.[47] These weapons posed a threat to the landing forces and naval vessels; they had to be eliminated. The Force was given the mission.

For the most part, the islands were beachless. Steep cliffs descended into the water in most areas, except for a few small areas on the northern side. Scrub-covered hills and some cultivated land characterized the interior. After a visual reconnaissance by submarine and rubber boat, Colonel Walker decided to assault the island from its cliff-strewn southern (seaward) side. Some French officers who knew the islands said such an approach was impossible, but they had not been at Difensa. The very difficulty of the seaward approach made it a logical choice for the Force.

The Force conducted six weeks of training at Santa Maria di Castellabate prior to the operation. All ranks received a refresher course in basic training. Recent replacements underwent instruction in the use of Force weapons and techniques. Physical fitness again drew a lot of emphasis to put the men in special combat form.

In August, the Force began intensive assault and amphibious training under the direction of the Invasion Training Center. The program included organizing boat teams and demolitions squads; wire breaching by boat teams; cliff scaling by day and night; use of rockets and flamethrowers; employment of waterproofing equipment; handling of mines and booby traps; swimming; training for infiltration; route marching; navigation; and landing techniques against a beachless shore.[48] Attached naval beach-marking parties and shore fire-control parties accompanied the Force during this training.

Several landing exercises also took place, including a night assault against two islands above Naples. Each regiment went through a full-scale dress rehearsal. Motorboats towed 10-man rubber boats to within 1,000 yards of the shore. Then, the boat teams paddled in, climbed the beach cliffs with full combat loads, prepared for counterattacks, and landed supplies. Critiques and corrective actions followed each exercise.[49]

On opening night, Force combat strength stood at 2,057 men. Its naval support force consisted of five transports, three medium and two small landing ships, a French battleship, one heavy cruiser, five light cruisers, three destroyers, sixteen PT boats, and fifteen small craft. (This naval force also supported a small French commando unit landing on the mainland.)[50]

The weather was ideal, the sea calm, and the night dark when the troop transports halted 8,000 yards offshore on the night of 14—15 August 1944. At the appointed time, the men of the Force climbed into their rubber boats and tied on to motorboats—three rubber boats to each tow line. Scouts in kayaks and electric surfboards marked the landing sites ahead of the main body. Shortly after midnight, the 1st Regiment landed 650 men on Port Cros, while the 2d and 3d Regiments (1,350 men) climbed ashore on Levant (see map 20).

Following the established Force pattern, the landing parties achieved complete surprise. Meeting no resistance, the men of the Force reached their assembly areas without trouble. On the Île du Levant, the Germans rapidly holed up in the port of Levant, where they fought hard against the 2d Regiment. By dawn, the Force had cleared a good beach, resupplied, and evacuated the wounded. By 2334 on 15 August, all resistance had ended. The coastal battery in the east turned out to be a dummy. At Port Cros, the German defenders held out for forty-eight hours. The last enemy strongpoint, sheltered in an old thick-walled fort, surrendered when twelve 15-inch shells from a supporting battleship passed from one side of the fort through the other. As in other Force operations, surprise, shock, tenacity, and leadership, were the key ingredients in the success.

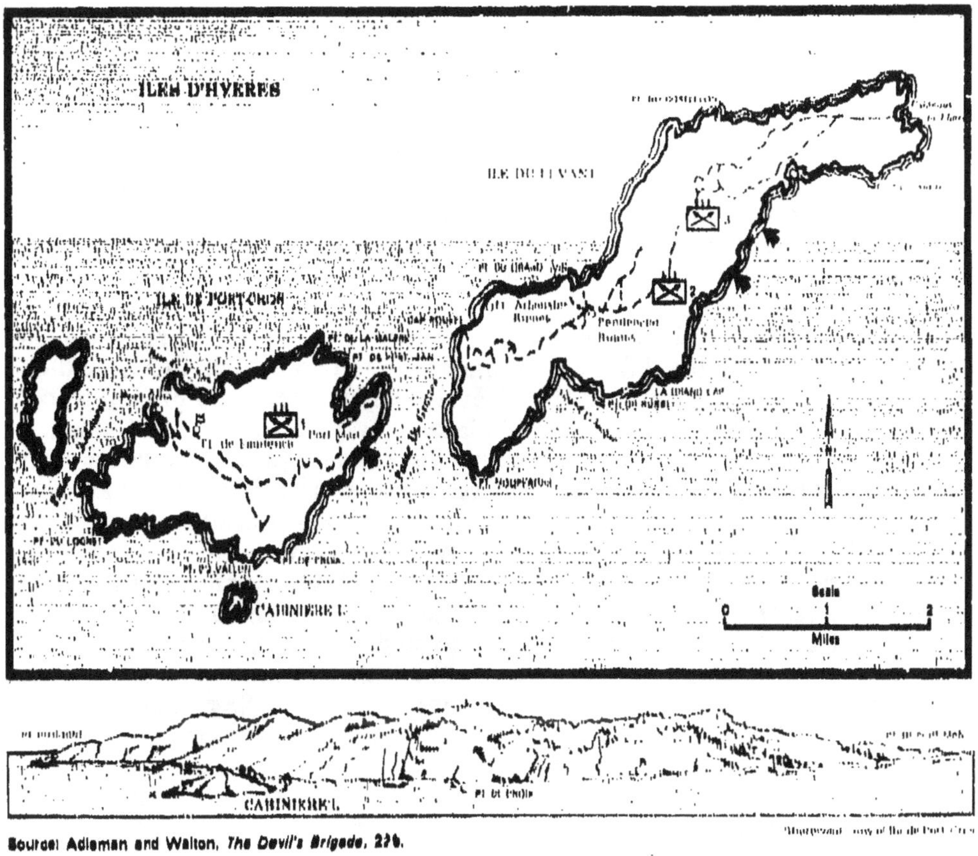

Source: Adleman and Walton, *The Devil's Brigade*, 226.

Map 20. Assault on Port Cros and Levant

The End of the Road

On completion of these missions, all three regiments of the Force were transferred to the French mainland, where they fell once again under General Frederick's command, the 1st Airborne Task Force (1st ABTF). After conducting a successful airborne assault during the invasion, the 1st ABTF was given the mission of advancing eastward along the Mediterranean coast to secure the rear and right flank of the main forces. The Canadian Army's report on its contingent in the FSSF succinctly describes this last phase of the Force's existence:

> There now began for units of the Force a series of rapid advances along the Mediterranean Coast that was to take them in less than three weeks a distance of some 45 miles to the Franco-Italian frontier. In general, enemy resistance was light. It was not necessary for the Force to mount any large-scale operations. Engagements were on a regimental, or lower, level, as the enemy fought small typical delaying actions. Each day brought its quota of two or three towns occupied, a number of machinegun positions destroyed, a score or so of PWs taken, a mined road crater filled or a bridge replaced by the Engineers. Casualties ... were slight. ... The greatest hardship on officers and men alike was the strain of being almost continually on the move, with no opportunity for rest or relaxation. ... It was not until November, when positions became stabilized on the frontier, that it was found possible to withdraw units into reserve.[61]

This entire advance was executed on foot without armor support. Two attached artillery battalions provided indirect fires.

The FSSF spent its last days sputtering into inactivity in defensive positions along the Italian border. The insignificance of its final operations contrasted starkly with the proud record it established in Italy, where it took the toughest tasks at the center of the action. But the Force was not completely forgotten. Since October, it had been the subject of message traffic and discussions between the U.S. War Department, the Canadian Department of National Defense, and the Mediterranean theater. The latter, had recommended the disbandment of the FSSF, a recommendation in which the Canadians concurred, primarily because of the difficulty which they were having in continuing to provide the Force high-grade infantry replacements. The Canadians now viewed the FSSF as an unproductive dispersion of scarce resources. Only the 6th Army Group lobbied for the retention of the FSSF because of its special capabilities in snow and mountain operations. Looking ahead to the advance into southern Germany, the army group envisioned the use of the FSSF in the Alps, then the French Vosges Mountains, and finally in the Black Mountains. The army group's concern was heightened by its recent rejection of the War Department's offer to it of the 10th Mountain Division (the army group rejected the offer because the 10th was not scheduled to arrive in theater until March 1945 due to a shortage of shipping space for pack animals). Lacking a proven mountain unit, the 6th Army Group wanted to hold on to the Force.[52] But the War Department turned down the army group's request. When the 1st ABTF was withdrawn from the Italian frontier in late November, the FSSF moved back to a holding area. On 5 December 1944, the color guard sheathed the Force colors during a final parade and memorial service.

The members of the Force were reassigned to other units. The older Canadian members, those with airborne training, joined the 1st Canadian Parachute Battalion in France. The larger remainder were shipped to Italy and fell into the general Canadian infantry replacement pool.[53] Edwin Walker, now a brigadier general, took some of the U.S. Force members with him when he assumed his new command, the 474th Infantry (Separate). At the war's end, the 474th shipped to Norway to oversee the repatriation of German soldiers who surrendered there.[54] The rest of the U.S. members of the Force finished their service in a variety of other units.

Tactical Style

The mountain tactics of the FSSF are outlined in detail in a Force report dated 14 April 1944 and titled "Lessons From the Italian Campaign." This report strongly emphasized that terrain had an overwhelming influence on the tactics employed by the Force.[55] "The most important lesson learned from the terrain was that without exception high ground must be taken and held."[56] To implement this principle, the FSSF insisted that movement to seize peaks take place as high up on the connecting ridges as possible without creating silhouettes on the skyline. Crests might be occupied temporarily in the defense, but over the long term, the reverse slopes were considered safer from enemy fires and observation. Hidden observation posts on the forward slope would then provide early warning to the reverse-slope defenders. In the attack, and even after a specific objective was taken, the Force found that it often received hostile fire from adjacent peaks and ridges. As a result, the Force carefully selected its routes of march and moved in the dark (including supply and evacuation) to avoid observation. Moreover, since mines were often laid in obvious tracks (particularly in draws and along ridgelines), mine detectors had to be employed forward, and well-worn trails went unused until cleared.

The tactical formations employed by the FSSF changed with the ground. The most common formation used was the wedge or arrowhead. This formation provided a heavy volume of fire to the front and the flanks, and it permitted quick transition into a skirmish line. Even veteran troops, however, showed a tendency to bunch up.

The platoon and section tactics practiced by the FSSF relied strictly on closing with the enemy as rapidly as possible. Using both fire and maneuver and fire and movement, Force units sought the flanks and rear of the enemy positions:

> Hit him hard and move in where our hand grenades are effective. He dislikes them.

> Aggressiveness and fast maneuvering to the flanks and, if possible, the rear we find the best policy for taking out machine gun nests. . . . Don't be pinned down by fire. . . . Quick thinking, maneuver fast is our policy.[57]

The men of the Force used a lot of hand grenades, particularly to reach into sheltered positions protected from direct fires. Personal accounts of tactical actions frequently praised the effectiveness of hand grenades in assaults.

Force units preferred to attack at night and repeatedly achieved surprise. During attacks, they discovered that the enemy usually did not aim his automatic weapons; the Germans fired along specific lines to cover areas of ground. Once they had determined these lines of fire, it was not difficult for the Force to move against the enemy flanks.

Control of movement and fire at night posed problems. The Force overcame these problems through five techniques: good training that simulated anticipated conditions; the use of SOPs; strict discipline; simple plans understood by all; and constant supervision. Local guides sometimes got lost, so leaders learned to rely on their own night navigation through prior map study, daylight reconnaissance, and frequent reference to maps and compasses during movement. Once contact was made at night, actions became automatic.

When attacking in daylight without surprise, the Force valued the support of armor, engineers, and artillery to reduce or suppress enemy strongpoints while they assaulted. They recognized, however, that the infantry played the decisive role:

> We have learned from experience, sometimes slightly bitter, a lesson of the utmost importance to infantry units and of particular importance to a force such as ours—the lesson of "self-reliance" by the employment of our own supporting weapons.

> It is very easy for subordinate commanders charged with the responsibility of the attack to overlook at certain phases of the attack the full employment of their own supporting weapons. Calls for artillery support are sometimes made where the task is one for the unit's mortars or heavy machine guns, if such are available. The rocket launcher and rifle grenade are not always fully exploited.[58]

When the FSSF engaged in active operations, it constantly sent out patrols. In the defense, as at Anzio, vigorous patrolling kept the enemy from laying mines and ambushes to his immediate front and prevented mortars from being set up too close. Force officers insisted that patrols have either the purpose of reconnaissance or raiding. Particularly at night, it was important to have a single objective for each patrol, one on which all patrol members had been fully briefed and rehearsed, if possible.

The Force believed reconnaissance patrols should be small, six to twelve men, armed with at least one light machine gun and several submachine guns. The FSSF learned in France that, in a fast-moving situation, early, aggressive reconnaissance yielded results not obtainable later. Such reconnaissance well forward of the advance of the main body often caught the enemy by surprise, before he had fully camouflaged his positions and hidden his troops.[59] Several Force company commanders recommended that each company specially train a number of small groups of soldiers for reconnaissance. The overall Force objective, however, was that *all* members of the Force be so trained. Reconnaissance by commanders was deemed to be especially important to the outcome of combat actions.

Combat patrols needed to be larger than reconnaissance patrols—up to platoon strength and with more automatic weapons. These patrols sometimes were split into two groups. One group accomplished the mission; the other group followed at some distance to prevent ambushes, to maneuver, if necessary, and to provide flexibility. To enhance control, the men of the Force preferred bright nights for raids and dark nights for reconnaissance.

As members of a light infantry force facing an enemy with substantial artillery and mortars, men of the Force learned that they had to dig in on the defense. Every time that the Germans lost a position to the Force, they immediately counterattacked with infantry, artillery, and mortars—compelling the Force to go to ground. One NCO in the 2d Regiment said, "When we are not shooting we are digging. When we are not digging we are shooting. When we take up a position, we dig and dig deep. It pays."[60]

The value of houses to the Force as outposts or defensive positions drew mixed opinions. While they afforded protection from fire, they could not be occupied carelessly; other local positions had to protect them. Men of the Force often viewed them as traps, objects to be blown up into rubble, not occupied.

In the rugged mountain fighting, choice of weapons and munitions often influenced the outcome of actions. Force leaders had to balance their desire to have heavy volumes of fire on a moment's notice with the burden of carrying heavy weapons and loads of ammunition by hand over mountain tracks. Still, they were loath to leave anything behind. Light machine guns and submachine guns were indispensable to the shock of a surprise assault by night. The Force often needed bazookas in each section to take out houses, bunkers, or an occasional light tank. Everyone used hand grenades. Unit mortars were often used to break up enemy counterattacks. The flamethrower, it appears, was the one weapon used in   ently. Heavier machine guns, .30 and .50 caliber, were used rimarily in the   nse at Anzio. All Force soldiers were expected to know how to operate all the available weapons.

Courtesy of Colonel Robert D. Burhans

Sergeant Cyril Krotzer of the 2d Regiment surveying the beachhead for the last time before preparing for the drive on Rome

Leadership

The FSSF's tactical style was unusual and distinctive, and its execution required exceptional leadership. Mountain warfare often demanded that junior leaders exercise their own judgment in the absence of detailed orders. The nature of the terrain broke up regimental and battalion attacks into connected series of section and platoon attacks. Under these decentralized conditions, aggressiveness and initiative—qualities possessed in abundance by Force leaders—were invaluable.

To ensure control, the Force wrestled with the problem of where commanders should be during attacks. Generally, the Force called for officers to lead from the front. However, if they were too far forward, commanders might be unable (or slow) to employ their reserves and control their support. The Force recognized that commanders' positions might vary with the situation, but it concluded that "in rugged mountainous country where maneuverability was limited, approaches long and time consuming, and communications sometimes unreliable, it was generally advisable [for commanders] to be with the forward troops."[61]

Another leadership problem recognized by the Force concerned the transmission of verbal orders. Force commanders insisted, as a rule, that all soldiers be briefed on tactical plans and expected enemy actions. Commanders attributed much of their tactical success to the fact that their soldiers knew the purposes of coming actions and how they fit into them. Mountain combat, however, often required changes in plans that needed to be disseminated rapidly by voice. Constant care was required to make sure that such orders (passed down by runner or relay) reached their proper destination.

During actual attacks, voice communication assumed great significance. A visible leader who calmly directed actions on the battlefield settled the men and spurred them on. A section leader in the 2d Regiment wrote: "Don't be afraid to talk loud if it's in the daytime, because your voice is a real tonic at this time, both to yourself and the men with you. Don't worry too much about how your men will conduct themselves when they get under fire, just keep yourself cool and your men will be right there acting like they had been under that stuff lots of times before."[62]

Elite soldiers, like the men of the Force, required that their leaders earn their respect. The concept of leading by virtue of rank was not recognized in the Force. The privilege of leadership fell only to those who had earned it through performance, first in training, then in combat. In practical terms, this meant that the men recognized the superior abilities of their leaders. Conversely, it meant that the Force leaders had to be as hardy, as fit, and as proficient in infantry skills as their men. Frederick's performance in this regard set a high standard for all his subordinates to emulate.

Logistics

Keeping support moving forward in the rugged Italian mountains was far from easy. Frederick frequently had to divert combat troops to act as porters. At Difensa, Frederick employed 1,200 men in this role, about 50 percent of his total manpower. The soldiers understood the necessity of such work; they jokingly nicknamed themselves "Freddy's Freighters." The demands of this labor, however, created many casualties—from exhaustion, enemy indirect fires, snipers, and exposure.

The command attempted to use aerial resupply in the mountains, but airdrops worked poorly. There was too much fog, and it was too easy to miss a drop site perched on some ridge or peak. Pack trains, on the other hand, helped enormously. The Force enjoyed such support in the Majo operations and from Anzio to Rome. The pack trains came from the Italian Army or from other U.S. divisions. The FSSF never received its own dedicated pack train.

208

The Force also never received its TOE authorization of vehicles. Consequently, when operating in more open terrain, men of the Force borrowed, scrounged, and stole vehicles whenever they could and used them for tactical transport and supply. At the rest area in Lake Albano, according to one source, the men stole so many vehicles for recreation that each soldier had his own personal vehicle.[63]

Frederick had an unusual attitude about rations in combat. At the front, he subsisted on instant coffee and cigarettes. Eating no food himself, he figured that his infantrymen could survive on two-thirds of a K ration per day. As a result, the FSSF carried few rations.[64]

Medical support to the Force appears to have been excellent. Aid stations routinely were established quite far forward, with evacuation being accomplished almost wholly by litter. Companies not in contact provided the manpower. Frederick, himself, pulled litter duty, demonstrating his own estimate of its importance. Force policy directed the immediate evacuation of casualties; holding casualties on the line tended to demoralize the men.[65]

During the course of its active operations, the Force established a reputation for lax supply discipline. On several occasions, they apparently discarded materials that they felt they no longer needed and did not want to continue to carry.[66] It appears that the Force tended to stock materials for the worst-possible eventualities. When actual situations required much less expenditure, the stocks were simply abandoned. The richness of the American supply system permitted such waste. When the supply system failed to come through, the Force improvised to correct any of its perceived deficiencies. Raiding German larders at Anzio, commandeering half-tracks en route to Rome, and filching vehicles at Lake Albano are three typical examples of their improvisations.

Finally, as an experiment in the division of manpower between combat and service-support functions, the Service Battalion proved to be a success. Aside from occasionally assigning it help from the combat echelon when necessary, Frederick apparently never meddled with its organization. No documented criticism of the unit exists.

Conclusions

Few units in World War II equaled the glowing reputation established by the FSSF. It never met defeat in battle. It accomplished the most difficult missions with an elan and a proficiency that astonished all outside observers, including the Germans. In size the equal of an infantry regiment, the Force consistently accepted tasks appropriate to a regular infantry division. Moreover, the unit remained effective even after it had sustained casualties that would have incapacitated another force. Yet viewing the entirety of its short history, not just its glories, one cannot ignore a number of problems internal and external to the Force. Both the reasons for the success of the Force and its problems deserve attention for the insights that they provide on current light infantry operations.

Certainly, the FSSF can be categorized as a light infantry force. It had no organic supporting arms. Its tactical mobility derived from its marching power and its ability to dominate the terrain. However, one must understand that the FSSF was more than a light infantry force. For a number of reasons, care should be taken in relating the employment and characteristics of the Force to other light infantry.

Because of the rigorous selection practiced by Frederick and his staff, the soldiers and officers accepted for training in the Force were of above-average quality. The intensity and mercilessness of the training further guaranteed that only the best of the initial cohorts would measure up. Together, rigorous selection and hellishly intensive training produced an extraordinary body of men who were closely knit, full of esprit, and confident. More highly trained than the U.S. Rangers, the Force was a true elite, a characterization that cannot be applied to all light infantry units.

The Force also differed from ordinary light infantry in that it spent almost a full year in training as a coherent organization before it was ever committed to battle. Probably the only other unit in the U.S. Army to enjoy such a prolonged, unified preparation was the 87th Mountain Infantry Regiment. The skills acquired by the Force in this training endowed it with an unmatched versatility. Exceptionally fit, the men of the Force possessed the abilities to ski, snowshoe, climb mountains, assault by parachute, destroy facilities with explosives, and conduct amphibious raids—all during the black of night. The average light infantry unit is proficient in some of these skills (and has cadres with experience in others), but they do not routinely train in them all.

As an elite Canadian-American outfit, the FSSF also had special access to resources not routinely available to regular infantry units. Highly placed persons kept tabs on where the Force was and what it needed. Frederick seemed to be able to obtain whatever materials and equipment that he wanted, changing his unit's TOE as he saw fit. The Force's abundance of light machine guns, submachine guns, bazookas, and mortars demonstrated its privileged access to resources. Frederick was also able to pick and choose the U.S. replacements for his casualties.

Any analysis of how and why the Force achieved so much success should be accompanied by a recognition of these features that distinguished it from ordinary light infantry. The Force's record should be viewed as a sort of upper limit to what light infantry can be expected to accomplish. The more common aspects of the Force training program—fitness training, marksmanship, night operations, and basic demolitions—and virtually all of the features of their tactical style, with its emphasis on speed, shock, surprise, aggressiveness, and terrain domination, can and should be accepted as legitimate models for light infantry units today. Moreover, the use of the Force as a flank-covering force in restricted terrain and their economy-of-force role in Anzio against relatively light enemy forces are appropriate examples demonstrating how light and heavy forces can be combined synergistically. (The use of the FSSF for these missions, however, wasted to a certain degree its one-of-a-kind offensive capabilities for deep raids.) Many of the special tasks accomplished by the Force could not have been performed by the average light infantry unit without specialized training (for example, its two nighttime amphibious raids and its assault against Difensa). The specialized training of the FSSF enabled it to be used as an operational-level, spearhead force in the invasions of Kiska and southern France.

The most significant problem experienced by the Force was its misuse by corps and army commanders. In this regard, the Force spent most of its existence performing missions that could have been accomplished as well by units with far less specialized training. The case can be made that only the Difensa assault and the amphibious strikes against Kiska and the Hyères Islands fully employed the special capabilities of the Force. The Majo operations might be added to the list. Nonetheless, the employment of the FSSF at Anzio, in the drive to Rome, and on the mainland of southern France failed to maximize its capabilities. Moreover, these operations embodied close to 90 percent of the time spent by the Force in contact with the enemy and accounted for the majority of its casualties.

The misuse of the Force conforms to a frequently observed historical pattern. Once a unit arrives in theater—its special capabilities notwithstanding—its availability irresistibly tempts commanders to employ it. Stilwell's poor use of the Chindits, the massacre of the Rangers at Cisterna, the frequent employment in Italy and France of airborne units as regular foot infantry, and Germany's use of its mountain divisions on the Russian steppes are other examples of this historical tendency.

The misuse of the Force may also have resulted because there were no legitimate special missions for it to accomplish. General Clark might not have been able to identify deep-raiding and reconnaissance missions appropriate for the Force. If so, though it seems unlikely, Clark understandably employed the FSSF rather than allow it to sit idly in the rear. Nevertheless, the Force never conducted a deep raid, performed deep reconnaissance, or used its oversnow and airborne capabilities. The time and resources it spent training for these capabilities was largely wasted.

Because of the relative scarcity of legitimate missions for specialized forces, the formation of such units should be limited. Most of the combat operations in a conventional war are of an ordinary sort and can be accomplished by normal units. When specialized operations are necessary, they can be undertaken by conventional units provided with special training prior to the operations.

To allow the proliferation of specialized forces in an army robs an army of its needed manpower and flexibility. To withhold such forces because the specific conditions appropriate to their employment have not appeared can be quite inefficient in the long run. A number of senior World War II commanders complained about the waste and futility of having so many specialized forces on hand.[67] Clearly, a correct balance must be struck between special and conventional forces.

Another problem connected with specialized forces and specifically with the FSSF was the question of replacements. To replace members of the Force with ordinary infantry was inappropriate. Thus, Canadians took care through their stateside FSSF establishment to send groups of replacements to the FSSF that had been specifically trained in typical Force skills. These replacement cadres and soldiers seldom fell short of Force expectations. However, this process was demanding and expensive. Eventually the Canadians tired of it and sought the disbandment of the Force.

No such pipeline for trained personnel existed for the Americans. Frederick maintained a relatively high standard of replacments by insisting on and receiving only the best men, but even so, it was impossible to get men as well trained as his original unit.[66] The FSSF was fortunate to receive the remnants of the Ranger units destroyed at Cisterna, since these men were much like the soldiers of the Force in terms of skills and attitudes. Inevitably, over a period of time, the overall quality of the Force deteriorated, although it retained its elite character to the end. The same deterioration occurred in airborne, mountain, and commando units on both sides during the war.

Several other unusual problems arose because of the unique composition of the Force. Because of its binational makeup, the Force was bounced around as a political football. Both the United States and Canada had to agree on where it was to be employed, and Winston Churchill was not above adding his influence to the equation, even though England had provided no men or materiel to the unit. The binationality of the Force created a number of problems to be reconciled: differences in pay, uniforms, decorations, promotions, punishments, and other administrative actions.

Another problem besetting the Force was its devil-may-care attitude, which, though it was encouraged in its training, led it into occasional overaggressiveness. The Force's tendency toward impatience and excessive bravado induced its leaders to expose themselves to excessive fire, for the men to advance ahead of flank units (as in the breakout at Anzio), and for units to attack fortifications rashly (as on Levant, without waiting for commanders to make up their minds about surrendering).

The Force also was troubled by a substantial number of prima-donna-type personalities. Highly individualistic and lacking the garrison discipline of a regular unit, the wilder element in the Force seemed always to be up to mischief when not involved in combat. These shenanigans included transporting prostitutes from Naples to Anzio, practicing hand-to-hand combat on unwary MPs, stealing jeeps, and using explosives for pranks. The leadership of the Force seemed to tolerate such antics as long as no one was seriously hurt, perhaps accepting the disorderly behavior as the price they had to pay for similarly bold performance in combat.

As a whole, the FSSF was a unique unit with exceptional capabilities quite beyond those of general-purpose light infantry organizations. While the FSSF should not be considered a prototype for the design or employment of subsequent light infantry units, nevertheless, a careful and judicious analysis of its training, leadership—and especially its tactical style—yields a number of important lessons applicable to light infantry. The misuse suffered by the Force also provides a warning to current light infantry units: there are no guarantees that senior commanders in the future will properly employ light infantry or exclude it from combat situations for which it is ill suited.

NOTES

Chapter 4

1. The exiled Norwegian government objected to the destruction of its hydroelectric stations. In addition, studies showed that the stations could be destroyed by strategic bombing; a ground raid was not required.

2. Canada, Department of National Defense, General Staff, Historical Section, "The 1st Canadian Special Service Battalion," Report no. 5 (N.p., n.d., mimeographed), 15, hereafter cited as Report no. 5. At various times, planners considered using the Force in the Caucasus, Balkans, North Africa, and Europe.

3. Robert D. Burhans, *The First Special Service Force: A War History of the North Americans, 1942—1944* (1947; reprint, Nashville, TN: Battery Press, 1981), 14, hereafter cited as Burhans, *FSSF*.

4. Report no. 5, 3.

5. General Paul D. Adams, Interview no. 1 with Colonel Irving Monclova and Lieutenant Colonel Marvin C. Lang, 5 May 1975, 64, photocopied transcript on file at the U.S. Army Military History Institute, Carlisle Barracks, PA, and at the Combined Arms Research Library, U.S. Army Command and General Staff College, Fort Leavenworth, KS.

6. Burhans, *FSSF*, 61.

7. Martin Blumenson, *Salerno to Cassino*, United States Army in World War II (Washington, DC: Office of the Chief of Military History, U.S. Army, 1969), 255.

8. Major General Robert T. Frederick, interview at the Pentagon, 7 January 1969, 6, photocopied transcript on file at the U.S. Army Military History Institute, Carlisle Barracks, PA, and at the Combined Arms Research Library, U.S. Army Command and General Staff College, Fort Leavenworth, KS. Also see Burhans, *FSSF*, 43.

9. Burhans, *FSSF*, 49. Ninety-nine percent of the men of the Force attained Norwegian Army standards in skiing within two weeks. See also Robert H. Adleman and George Walton, *The Devil's Brigade* (Philadelphia, PA: Chilton Books, 1966), 84. The culmination of the ski training was a thirty-mile cross-country ski trip with full pack and rifle.

10. Adams interview, 62; and Adleman and Walton, *Devil's Brigade*, 84.

11. Report no. 5, 12.

12. Burhans, *FSSF*, 23.

13 Ibid., 24.

14. Adams interview, 60; and Adleman and Walton, *Devil's Brigade*, 69, 78—79.

15. Ibid., 58.

16. Burhans, *FSSF*, 44.

17. Adams interview, 60.

18. Adleman and Walton, *Devil's Brigade*, 93; and Burhans, *FSSF*, 56.

19. Burhans, *FSSF*, 58.

20. Ibid., 83; and Report no. 5, 29—30.

21. Chester D. Starr, ed., *From Salerno to the Alps: A History of the Fifth Army, 1943—1945* (Washington, DC: Infantry Journal Press, 1948), 52—53. See also Blumenson, *Salerno to Cassino*.

22. Blumenson, *Salerno to Cassino*, 265. See also U.S. Military Academy, West Point, Department of Military Art and Engineering, *Operations in Sicily and Italy, July 1943 to December 1944* (West Point, NY, 1950), 53—54; and Fridolin von Senger und Etterlin, *Neither Fear Nor Hope* (New York: E. P. Dutton and Co., 1964), 184. Senger und Etterlin states that the German positions were vulnerable because they lacked the manpower to man the entire line.

23. U.S. Army, Special Staff, Historical Division, "Critical Evaluation of Italian Campaign Based upon German Operational and Tactical Viewpoints, 3 September 1943—2 May 1945," (N.d.), 44, hereafter cited as "Critical Evaluation"; and Senger und Etterlin, *Neither Fear Nor Hope*, 184.

24. Blumenson, *Salerno to Cassino*, 265.

25. "Critical Evaluation," 44; and Senger und Etterlin, *Neither Fear Nor Hope*, 187.

26. Adleman and Walton, *Devil's Brigade*, 126—27.

27. Ibid., 137; and Burhans, *FSSF*, 111—12.

28. Burhans, *FSSF*, 124.

29. Ibid., 120.

30. Adleman and Walton, *Devil's Brigade*, 133.

31. "General Frederick and His North Americans," *Reader's Digest*, 45 (November 1944):101.

32. Burhans, *FSSF*, 145.

33. Ibid.

34. Adleman and Walton, *Devil's Brigade*, 160.

35. Ibid., 157.

36. U.S. War Department, General Staff, *Fifth Army at the Winter Line, (15 November 1943—15 January 1944)*, American Forces in Action Series (Washington, DC: Military Intelligence Division, U.S. War Department, 1945), 98, hereafter cited as WD, GS, *Winter Line*.

37. Ibid., 99; and Burhans, *FSSF*, 154.

38. Report no. 5, 38.

39. First Special Service Force, "Lessons from the Italian Campaign" (14 April 1944), 2, hereafter cited as 1st SSF, "Lessons."

40. Burhans, *FSSF*, 170.

41. Ibid., 179.

42. Ibid., 185.

43. Adleman and Walton, *Devil's Brigade*, 177.

44. Burhans, *FSSF*, 209; and Report no. 5, 42.

45. Frederick interview, 2.

46. Ibid., 15.

47. U.S. Army, 7th Army, *History of the Seventh Army*, Phase 1 (N.p., 1945), 60.

48. Ibid., 121—22; and Report no. 5, 46.

49. Ibid., 122.

50. Ibid., 32. The FSSF and its supporting units were called the Sitka Force for this operation.

51. Report no. 5, 48.

52. U.S. Army, 6th Army Group, *History of Sixth Army Group, July—October 1944*, vol. 1 (N.p., 22 June 1945), 185.

53. Report no. 5, 52.

54. Adleman and Walton, *Devil's Brigade*, 244.

55. 1st SSF, "Lessons." In addition, the following reports include some lessons learned by the FSSF and the 1st Airborne Task Force during the campaign in southern France: U.S. Army Ground Forces Board, Mediterranean Theater of Operations, "Report on Airborne Operations in DRAGOON," by Colonel Paul N. Starlings, Report no. A-200, 4 November 1944; and U.S. Army Ground Forces Board, Mediterranean Theater of Operations, Report no. A-228, 1944?, hereafter cited as AGFB, MTO, Rept. no. A-228.

56. 1st SSF, "Lessons," 1.

57. Ibid., 4—5.

58. Ibid., 5.

59. AGFB, MTO, Rept. no. A-228, "Lessons Learned in Operations from 15 August 1944 to 1 November 1944," 5, submitted by Colonel H. T. Brotherton as enclosure 3.

60. 1st SSF, "Lessons," 5.

61. Ibid., 2.

62. Ibid., 8.

63. Adleman and Walton, *Devil's Brigade*, 219—20.

64. Frederick interview, 16.

65. 1st SSF, "Lessons," 15.

66. Report no. 5, 28—29, 48.

67. Roger A. Beaumont, *Military Elites* (Indianapolis, IN: Bobbs-Merrill Co., 1974), 8. At the end of the war, the U.S. Forces European Theater Headquarters convened a board of general officers to record important combat lessons. One of the recommendations of this board was that no specialized divisions be retained in the force structure except airborne divisions.

68. Frederick interview, 4.

BIBLIOGRAPHY

Chapter 4

Adams, Paul D., General. Interview no. 1 with Colonel Irving Monclova and Lieutenant Colonel Marvin C. Lang, 5 May 1975. Photocopied transcript on file at the U.S. Army Military History Institute, Carlisle Barracks, PA, and at the Combined Arms Research Library, U.S. Army Command and General Staff College, Fort Leavenworth, KS.

Adleman, Robert H., and George Walton. *The Devil's Brigade.* Philadelphia, PA: Chilton Books, 1966.

Beaumont, Roger A. *Military Elites.* Indianapolis, IN: Bobbs-Merrill Co., 1974.

"The Black Devils.'" *Time* 44 (4 September 1944):63—65.

Blumenson, Martin. *Salerno to Cassino.* United States Army in World War II. Washington, DC: Office of the Chief of Military History, U.S. Army, 1969.

Burhans, Robert D. *The First Special Service Force: A War History of the North Americans, 1942—1944.* 1947. Reprint. Nashville, TN: Battery Press, 1981.

Canada. Department of National Defense. General Staff. Historical Section. "The 1st Canadian Special Service Battalion." Report no. 5. N.p., n.d. Mimeographed. Contains some loose, uncataloged material.

First Special Service Force. "Lessons from the Italian Campaign." 14 April 1944.

Fisher, Ernest F. *Cassino to the Alps.* United States Army in World War II. Washington, DC: Center of Military History, U.S. Army, 1977.

Forty, George. *Fifth Army at War.* New York: Charles Scribner's Sons, 1980.

Frederick, Robert T., Major General. Interview at the Pentagon, 7 January 1949. Photocopied transcript on file at the U.S. Army Military History Institute, Carlisle Barracks, PA, and at the Combined Arms Research Library, U.S. Army Command and General Staff College, Fort Leavenworth, KS.

"General Frederick and His North Americans." *Reader's Digest* 45 (November 1944):99—102.

Jackson, William Godfrey Fothergill. *The Battle for Rome.* New York: Charles Scribner's Sons, 1969.

"Monte la Difensa." Battle Analysis by Staff Group 14-A. U.S. Army Command and General Staff College, Fort Leavenworth, KS, May 1984.

Senger und Etterlin, Fridolin von. *Neither Fear Nor Hope.* New York: E. P. Dutton and Co., 1964.

Starr, Chester G., ed. *From Salerno to the Alps: A History of the Fifth Army, 1943—1945.* Washington, DC: Infantry Journal Press, 1948.

U.S. Army. 1st Airborne Task Force. "General Summary of Operations." N.d.

U.S. Army. 7th Army. "Field Order no. 1 (ANVIL)." 29 July 1944.

————. "Field Order no. 2," 19 August 1944.

————. *History of the Seventh Army.* Phase 1. N.p., 1945.

U.S. Army. 6th Army Group. *History of Sixth Army Group, July—October 1944.* Vol. 1. 22 June 1945.

U.S. Army Ground Forces Board. Mediterranean Theater of Operations. "Lessons Learned in Operations from 15 August 1944 to 1 November 1944." Submitted by Colonel H. T. Brotherton as enclosure 3 to Report no. A-228. 1944?

————. "Report on Airborne Operations in DRAGOON." By Colonel Paul N. Starlings. Report no. A-200. 4 November 1944.

U.S. Army. Special Staff. Historical Division. "Critical Evaluation of Italian Campaign Based upon German Operational and Tactical Viewpoints, 3 September 1943—2 May 1945."

U.S. Military Academy, West Point. Department of Military Art and Engineering. *Operations in Sicily and Italy, July 1943 to December 1944.* West Point, NY, 1950.

U.S. War Department. General Staff. *Anzio Beachhead (22 January—25 May 1944).* American Forces in Action Series. Washington, DC, 1947.

————. *Fifth Army at the Winter Line (15 November 1943—15 January 1944).* American Forces in Action Series. Washington, DC: Military Intelligence Division, U.S. War Department, June 1945.

Chapter 5

The Nature of Light Infantry

The light infantry forces discussed in the preceding four case studies vary widely from each other in several ways: size, organization, assigned missions, nature of the threat, and area of employment. In spite of these differences, a large number of elements are shared by light infantry forces—elements that describe what may be termed "generic light infantry." These distinguishing characteristics are not exclusive to light infantry; historically, other infantry units have displayed many of the same qualities. The separation line between conventional and light infantry is blurred; an overlap exists. Nonetheless, light infantry has exhibited these characteristics more uniformly and to a much greater degree than other infantry organizations.

General Characteristics

There are four primary characteristics that distinguish light infantry forces from regular (dismounted, motorized, or mechanized) infantry. The most important of these characteristics is an attitude of self-reliance. Self-reliance forms the essence of the light infantry ethic, the fountainhead from which all of its other characteristics flow. This attitude of self-reliance is exhibited by light infantry forces in a number of ways. For example, light infantrymen typically demonstrate strong confidence that they will survive and succeed in whatever situations they are found. They are undaunted by unfavorable conditions (such as being cut off or outnumbered). Their resourcefulness permits them to devise schemes to accomplish their missions, no matter how difficult the tasks. Furthermore, light infantrymen are accustomed to austerity. They have learned to do without comforts and benefits that other soldiers consider to be necessities. They are not psychologically tied to a logistic lifeline. Their attitude of self-reliance leads them to use any available resource to sustain themselves or to improve their combat capabilities. Moreover, light infantrymen do not give up. Even when outcomes seem inevitable, light infantrymen stay in the fight and attempt to turn situations to their advantage. Their self-reliance is typified by self-denial, fortitude, tenacity, and resourcefulness.

This attitude of self-reliance gives light infantrymen a psychological advantage over their enemies. Confident in their abilities, light infantrymen normally consider themselves to be tactically superior to their opponents. Once they have demonstrated this tactical superiority, their enemies often become fearful and wary. Light infantrymen use this psychological advantage to keep their enemies off-balance and tense. Unpredictable, invisible to view, employing methods not anticipated by their enemies, light infantry forces can often paralyze the minds and wills of their enemies before a battle begins.

219

This self-reliant attitude enables light infantry units to become the masters of their environment. Light infantrymen do not fight, fear, or resist the environment; they embrace it as shelter, protection, provider, and home. They learn to be comfortable and secure in any terrain and climate, be it jungle, mountain, desert, swamp, or arctic tundra. Exceptionally adaptable, light infantry units dominate the terrain in which they operate and use it to their advantage against their enemies.

As a result, light infantry forces exhibit a well-developed appreciation for the tactical aspects of ground. Because they understand and accept the terrain and climate as their natural environment, light infantry forces possess an unmatched tactical mobility on difficult ground. Moving with a speed and ease that astounds, light infantrymen routinely use routes and traverse areas deemed impassable by regular troops. Naturally, this terrain specialization takes time to develop.

Mastery of the environment and the attitude of self-reliance give the light infantry an unusual versatility. Light units adapt quickly from one environment to another or from one type of operation to another. Abrupt changes in plans find them still ready for action. Holding a jungle base one day, they may be ordered to conduct a deep raid, mount a long-term reconnaissance patrol, participate in a riverine operation, or attack a fortified position on the next. In addition, they can operate independently or in conjunction with larger forces. They can also function with or without significant combat support. Unexpected situations do not throw them off-balance. With additional specialized training (for example, airborne training), light forces can become even more versatile.

Their versatility is also reflected in a propensity for improvisation and innovation. Light infantrymen naturally derive new tactics, if necessary, because they are not tied dogmatically to a specific doctrine. They use their equipment in innovative fashion, and they do not hesitate to use the enemy's weapons and resources when they can. They also remain open to new ideas, new technology, and new weaponry. Light infantry forces maintain a flexible attitude toward the battlefield.

Given their dynamic characteristics, is it any wonder that light infantry forces typically possess high esprit. They know that they are different. They are proud of their ability to operate in the most difficult terrain, and they know that they are often assigned the most demanding missions. Confident and secure in the awaren.   of their unique tactical skills, light infantrymen consider themselves to be a cut above the average soldier. However, the general characteristics of light infantry do not appear automatically; they are developed through training, enlightened leadership, and actual operations.

Selection, Organization, and Training

Unlike highly specialized light infantry units like the Rangers, FSSF, and SAS, most light infantry forces are not composed of elite troops. Nonetheless, given excellent leadership and a demanding training program, light infantry units can develop an elite character. As they are pushed to standards of performance that seem out of their initial reach, light infantry soldiers can acquire the sense that they are something special. The development of the Golani Brigade in the Israeli Defense Force is a perfect example of this process. Although that brigade is composed primarily of conscripted soldiers, it is considered, because of its high standards of performance, to be one of Israel's finest. The brigade's standards are based on difficult training, exacting leadership, and the expectation that it will be employed in close terrain on the toughest missions.[1] The Golani Brigade, as well as the Chindits and the CCF, illustrate that light infantry soldiers need not be selected through a special screening process.

Light infantry forces also need not conform to a standard organization. Light infantry units may even organize internally in different ways. Organizers of such units, above all, seek forces with tactical flexibility. In this regard, the 3x3 squad organization found in the CCF in Korea and the British forces in Malaya and Borneo bears close scrutiny. This organization provides squad leaders with three discrete elements that they can use for reconnaissance, security, maneuver, or fire support as they see fit. Certainly, it gives squad leaders more options than the standard two fire team organization. However, in a scenario where staying power and conventional tactics are likely to be employed (for example, as with the FSSF in Italy), a standard nine- to twelve-man squad of two fire teams may be preferred.

Light infantry forces typically possess little heavy equipment and transport. However, they often acquire such support temporarily when needed. The bulk of light infantry firepower at the battalion level and lower is self-generated with light and medium machine guns, 2-inch to 81-mm mortars, rocket launchers, automatic rifles, hand and rifle grenades, and individual weapons. At a higher level, it is not unusual for light infantry organizations to be supported by artillery, light armor, tank destroyers-assault guns, engineers, aviation, and air forces. In general, however, light infantrymen tend to focus on the effective use of their own organic infantry weapons (a manifestation of their self-reliance). If equipment cannot be man or mule packed, the light infantry often has no use for it.

A number of common themes dominate light infantry training. For one, light infantry forces train under austere conditions. Comfort and luxury are unknown to them. Misery and privation prevail. Light infantrymen are taught to be self-reliant by being denied the things that they think they need in training. Food, water, rest, shelter from the elements: all of these fundamental needs are cut to the bone during light infantry training. Light infantry soldiers are pushed to the limits that they think they can endure—and then beyond. If they do not break, they learn that they can do things they never imagined they could and that they can continue to perform even though they are miserable and exhausted. Their capabilities are thus stretched. This austere, demanding training ultimately produces high self-confidence, trust, and cohesion within light infantry units. Light infantrymen often find that combat conditions are actually less severe than the conditions they experienced in training.

In addition to austerity and rigor, light infantry training puts great emphasis on physical fitness. Light infantry operations almost always place physical demands on soldiers far in excess of those endured by regular infantry. Conspicuous examples are the experiences of the FSSF paratroops at Difensa, Galahad at Myitkyina, the CCF in Korea, and British paratroops in the Falkland Islands. Thus, physical fitness training is integrated continuously into light infantry training. Troops do not ride to the rifle range; they march or run with weapons. Long marches with full rucksacks are commonplace. Twelve- to eighteen-hour training days develop endurance. Then, as competition grows, standards are raised. In the process, the old, the infirm, and the mentally weak are purged. The ultimate goal is to develop deep reservoirs of strength and stamina in the men. However, mere endurance is not enough; light infantrymen must also remain observant, alert, and ready for action while they are under physical stress.

Another theme of training is the development of initiative, particularly for NCOs and junior officers. Initiative and flexibility are developed by introducing unanticipated requirements into the training and requiring a response. A further technique is to place the burden of responsibility for some of the training on these junior leaders, requiring them, sometimes on short notice, to produce a plan and obtain the resources necessary for the training. Small-unit tactical exercises, such as patrolling and infiltration, also develop initiative. Like physical fitness, the development of initiative is integrated into the training program wherever possible. It is clear that these three training themes—austerity, physical endurance, and initiative—contribute directly to the development of the four primary characteristics of light infantry forces described earlier.

All light infantries seem to focus on several common skills in their training programs. Expert marksmanship, for example, is cited constantly as a fundamental skill. While all infantrymen—indeed all soldiers—must know how to shoot, light infantry units approach marksmanship as an art. Moreover, light infantrymen must have detailed knowledge of *all* the infantry weapons in their company, including crew-served machine guns. They spend hours on the range, day and night, refining accuracy and speed in all kinds of weather and simulated combat conditions. This training usually includes a heavy dose of maintenance training, actual practice on enemy weapons, and marksmanship competition internal to the unit. Familiarity with weapons has no relevance or meaning to light infantrymen: the achievement of expert-level skills is their goal. Light infantrymen are weapons masters.

Light infantry training emphasizes a variety of other skills and abilities: pioneer skills (to reinforce and exploit the terrain); the use of explosives; high standards of land navigation; hand-to-hand combat; field craft, small-unit tactics tailored to the operational environment; cross-training to spread expertise in a number of special skills (for example, artillery observation, communications, and mortar fire); and stealth. In addition, light infantries normally receive some form of specialized training to permit them to operate in unusual environments. Thus, all British infantry battalions passed through the Jungle Warfare School before their actual employment in Malaya and Borneo. The Chindits trained intensively in river-crossing operations, and the FSSF learned to ski because of the uniqueness of their intended employment. Light infantry forces also train extensively at night. In fact, light infantry views the nighttime as its natural period of activity.

Operations and Tactics

Light infantry forces are usually employed against other light forces at night and in close terrain, that is, terrain that restricts easy movement by heavy, mechanized forces. Light forces cannot survive in open terrain against heavy forces, although they may be used in open terrain against enemy light forces.

Light infantry forces tend to hide and rest during the day and to move and fight at night. The vulnerability of light infantrymen to enemy artillery and air compels them to use the cover of darkness for protection. The exception is jungle warfare. Thick jungle provides good protection from observation during daylight. At night, the jungle can sometimes be so dark that only limited movement is advisable. In the main, however, nighttime is preferred by light forces. Violations of this practice often lead to heavy casualties for light infantry in combat.

Light infantry forces are best suited for offensive operations. Indeed, the very character of the light infantry is to be offensively oriented, to retain the initiative in combat. Constantly probing, pushing, and challenging the enemy, light infantry forces cause the enemy to react to their activity, not vice versa. Even when employed in an overall defensive strategy (such as the FSSF at Anzio and Walker's forces in Borneo), light infantry constantly seeks opportunities to conduct offensive operations.

In the offensive role, light forces can be used at the operational level of war, as the Chindits were, although such historical examples are rare. Employing light forces at the operational level of war, however, usually requires a "break-in" capability, such as airborne training (for example, the German conquest of Crete) or amphibious training (for example, the FSSF at Kiska and in southern France).[2]

Although best suited for offensive tasks, light infantry forces have been used occasionally in essentially static defensive roles. But such missions fail to capitalize on the special capabilities of light infantry, particularly their superior tactical mobility, stealth, and offensive attitude. Tying light infantry units down in such a manner also increases their vulnerability to enemy fire. Whenever possible, commanders should seek to use light infantry units in offensive roles, even when the main conventional forces are on the defensive. In particular, commanders should be careful about using light infantry units as isolated strongpoints because of their lack of heavy weapons and their deficiency in staying power.

Brigade- and battalion-level operations are rarely conducted using light infantry forces, except in those instances where entire armies are organized on a light infantry basis (for example, the CCF and also the Japanese Army in World War II). Even when light infantry forces have been organized into divisions and brigades, their actual operations have tended to be extremely decentralized. The Chindits were organized into brigades and battalions, but the normal fighting organizations were the columns (i.e., half battalions). The FSSF, also organized as a brigade, most often conducted its attacks against company and battalion objectives, but subunits in the FSSF were small (companies numbered 100 men, battalions 300 men). British operations in

Malaya and Borneo rarely exceeded battalion level. While there were exceptions to this practice of employing relatively small units, most exceptions occurred when light infantry units were directed to conduct more conventional operations, such as the FSSF's breakout from the Anzio beachhead, the Chindits' permanent block on Japanese lines of communication in Burma, and the 77th Chindit Brigade's attack on Mogaung.

In general, companies, platoons, and squads of light infantry do the fighting—as a rule, in isolated actions. One reason for this circumstance is that close terrain tends to fragment battle into separate small-unit actions. Another is that light infantry forces often are required to operate in wide expanses of territory, leading commanders to divide their forces into small packets to cover the zone. (This phenomenon is a typical feature of economy-of-force roles.) A final reason for the use of small light infantry forces is that when the forces are used in a raiding or reconnaissance role, their objectives are usually company size and smaller. Because light infantry is most often used in decentralized, small-unit actions, light infantry trainers should devote the majority of their unit field training to company-level and lower tactics.

Light infantry forces also appear to operate frequently in conjunction with native irregulars (for example, the Kachins in Burma, and Ibans in Borneo) and with special operations forces (SOF), particularly in low-intensity conflicts. Interestingly, both the SOF and the native irregulars provide the same aid to the light infantry—intelligence, early warning, and security—even though they may not be working together. The light infantry, in turn, provides combat power when needed to the local forces and SOF.

Finally, light infantry operations are conducted at very close range. Light infantrymen normally do not seek to maximize the range of their weapons. Instead, they seek to get close enough to the enemy to smell and hear him.

The operational parameters described above dictate a unique tactical style for light infantry forces. The conventional tactics practiced by regular infantry forces and characterized by artillery preparations, significant combat support, massing of combat power, and large-unit maneuver do not work well for light infantry. Instead, light infantry tactics are characterized by three main features: surprise, shock, and speed.

Light infantry achieves surprise in both time and space through several means. Through superior field craft and domination of the terrain, light infantrymen approach enemy positions with animal-like stealth. Moving at night, using every fold in the ground, exploiting every bit of concealment, and making no noise, light infantrymen frequently reach hand-grenade range of the enemy positions before they are detected. Light infantry also attacks from unexpected directions and from more than one direction when feasible. Through preattack reconnaissance, light infantry leaders determine the weaknesses and gaps in enemy dispositions, which then become the objects of attacks. Finally, light infantry forces vary the time and style of their operations. Thus, the enemy is unable to predict their actions. (Consider, for example, how the British used rainstorms in Malaya to cover their approach for an attack against the Communist forces.)

Having achieved surprise, light infantry forces shock the enemy with the speed and power of their attack. Although lightly armed, light infantry can deliver a heavy volume of fire for short periods of time by massing all its weapons forward in coordinated, accurate fire. The application of such heavy firepower (especially of light automatic weapons), combined with rapid maneuver to the flanks and rear of the enemy's positions, creates a violent shock effect that can lead to quick victory. Light infantrymen must avoid being pinned down in the attack. Once pinned down, they lose momentum, their shock dissipates, and the enemy can introduce his indirect-fire weapons or reinforce his positions to the great disadvantage of the light infantry. Since light infantry often experiences problems with its supply of ammunition, it favors a quick decision in battle. As a result, light infantry units will some-times risk high casualties by storming a position, rather than allowing a lengthy battle to continue.

Light infantry units also exploit speed in their operations. To achieve speed, which is a function of superior individual and group tactical movement, light infantry relies on its intimate knowledge of the terrain, a high level of fitness, expert field craft, and the capability to negotiate difficult ground. By moving to an objective faster than its enemy thinks possible, light infantry can achieve surprise. To execute surprise successfully, however, requires stealth. While many units can only move rapidly if they make no attempt to conceal their movements, light infantry must accomplish stealth as well as speed if it is to be effective. Therefore, light infantry units keep to the wilds and seldom use roads or trails.

Decision making in the light infantry is also characterized by speed. Light infantry leaders must be prepared to react immediately to changes in tactical plans or to unforeseen situations. They seldom have the luxury of other forces coming to their rescue. They cannot afford to be pinned down mentally. For them, snap decisions often mean the difference between success or failure. Delaying a decision is usually dangerous.

The most frequently conducted tactical operation in the light infantry repertoire is patrolling. Patrolling requires high levels of discipline, patience, stamina, and field craft to be effective. Furthermore, patrolling techniques must be tailored to the specific tactical environment. True expertise is achieved only after constant practice in training and operations. Patrolling is probably the most mentally and physically stressful of all light infantry operations, because it involves extended exposure to the enemy, whose dispositions are not known, and it requires constant, all-around security. If a light infantry unit can master patrolling, its other operations will come much easier. Conse-quently, patrolling (particularly long-term patrolling) should be a staple in the light infantry training program.

During low-intensity operations, patrols may stay out for extended periods of time. Moreover, saturation of patrolling by light infantry forces has proven to be probably the most effective tactic of counterinsurgency. When the enemy is located through saturation patrolling, light infantry units typically conduct relentless pursuit to destroy him.

Another common feature of light infantry tactics is reliance on SOPs. These SOPs direct rapid, immediate actions under certain predictable situa-tions, such as crossing a stream, entering an ambush, locating an enemy position, encountering a moving enemy unit, stopping for a rest halt, or

225

relieving another unit. Light infantry, in such instances, operates with a minimum of orders. SOPs are never doctrinaire. When the tactical situation changes, SOPs change too.

In addition to preattack reconnaissance, light infantry units also conduct rehearsals as a matter of course. Rehearsals are made as realistic as possible and are conducted over the same type of terrain and under the same conditions as the actual attack. Rehearsals for night operations take place at night.

When used for deep tactical operations, light infantry forces often infiltrate at night on foot. Infiltration is conducted by small parties, at staggered times, through several seams in the enemy lines. Such small groups reconsolidate in the enemy's rear to accomplish the assigned mission. The CCF excelled in this technique.

The Defense

When light infantry is employed in the defense, its tactical mobility is impaired; it is vulnerable to artillery and aircraft fires; and it lacks the necessary staying power and firepower often needed in the defense.

Light forces defend best in depth, where they have more room to maneuver. Such maneuver provides them protection (to avoid enemy fires) and permits them to engage in limited offensive actions, such as spoiling attacks or defensive attacks against the flanks of an enemy force in motion. Maneuver space also allows light infantry to avoid being fixed and destroyed in detail.

If light forces are employed in a positional (static) defense, they must have time to dig strong fortifications characterized by overhead cover, bunkers, deep trenches, and switch positions that use the natural strengths of the terrain. In addition, they must be augmented extensively with combat support and logistical support: artillery, heavy machine guns, antiarmor weapons, barrier materials, and ammunition stocks. Only then will light infantry have the staying power and the survivability to be effective. Failure to take these precautions may result in the loss of the force.

In a positional defense, light infantry forces often defend from the reverse slope, placing only observers and limited firepower on the forward slope (to force the enemy to deploy early). Indirect fires attack the enemy on the forward slope and crest. Then, light infantry's organic, direct-fire weapons engage the enemy as he reaches the crest and begins to descend. Here, the enemy no longer enjoys the support of his own indirect and heavy direct-fire weapons and becomes vulnerable to surprise fires coming from a multitude of camouflaged positions that he has not been able to locate beforehand. Reverse-slope positions are also less vulnerable to artillery and air attacks.[3]

Camouflage is very important in the defense; one might term it an imperative. Expert camouflage, using natural and man-made materials, permits the light infantry to avoid detection and attacks by artillery and air observers. Light infantry forces in the defense should be invisible in the daytime. Repositioning and resupply efforts should occur at night.

Light infantry units show special skill in the coordinated use of machine guns and mortars in the defense. This skill should not be dismissed lightly, because it requires a refined appreciation of the military aspects of terrain and comprehensive experience in the siting of weapons, control of firing, and choice of targets.

Finally, light infantry typically conducts counterattacks as an integral part of its defensive scheme. Counterattacks take several forms. Light forces may attack to disrupt an enemy force forming for its own attack. They may attack the flanks or rear of an enemy force already deployed and in motion. Light infantry units also counterattack immediately to recover lost positions. Counterattacking forces may be held in sheltered positions to await specific opportunities, or they may be formed on the spot from available forces. Surprise and shock characterize such counterattacks.

Once attack objectives are secured, light infantrymen dig in rapidly to defend against the enemy's own counterattacks and to survive his artillery fires. In any temporary defense, light infantrymen dig quickly and deeply to provide themselves protection.

Combat Support

Although light infantry traditionally depends on its own organic fires, when placed in the defense or given conventional offensive missions, it must have additional combat support. Common sense and the factors of METT-T (mission, enemy, terrain, troops, and time available) dictate the scale and nature of the support. The support given the FSSF in Italy serves as an example of a properly supported force, while the experiences of Galahad at Myitkyina and the Chindits at Mogaung illustrate the consequences of failing to provide needed support.

Strong artillery support is essential for light infantry in the defense to reduce the combat power of the enemy before he closes to rifle range. In an attack against fortified positions, artillery is essential for light infantry to reduce enemy strongpoints, to force him to keep his head down, and to enable the light infantry to get close enough to be effective. Close air support complements the artillery.

Engineer support also increases the strength of the light infantry's defense through survivability (the construction of fortifications) and through counter-mobility (the erection of barriers and obstacles). The more engineer support available, the less time the light infantry needs to prepare to defend.

In addition, light infantry forces can frequently benefit greatly through the use of light armor because of its mobility, antiarmor weaponry, and capability to destroy hard point targets, such as enemy bunkers. Even in close terrain, light armor has proven valuable. Several modern light infantries include light armor in their force structure.[4]

Lastly, the helicopter has had an almost revolutionary effect on certain aspects of light infantry operations, particularly tactical mobility and resupply. In Borneo, helicopters compensated for several severe tactical disadvantages suffered by the British infantry and enabled it to react immediately to Indonesian raids. Helicopters neutralized the enemy's freedom of movement. Resupply by helicopters also permitted the infantry patrols and base camps to remain deployed for long periods of time. (The light planes used in the Chindit campaign performed similar functions and can be viewed as precursors to the helicopters.) In addition, the use of helicopters for deep insertion, casualty evacuation, and command and control should not be overlooked. No current light infantry force should be without helicopter support.

Leadership

Light infantry forces are often led at the highest level by bold-thinking, charismatic men. Wingate, Frederick, Templer, and Walker were extraordinary men of unusual talent, imagination, vision, and perseverance. Other examples outside the case studies examined here are David Stirling, founder of the SAS; General Karl Student, leader of the German airborne corps; T. E. Lawrence ("Lawrence of Arabia"); and Colonel (posthumously Brigadier General) William O. Darby, leader of American Rangers. While these officers were unorthodox in varying degrees, many of them succeeded equally well as leaders of conventional units. The binding thread in their characters appears to be their willingness to implement (and indeed to produce) innovative and unorthodox ideas about combat. Light infantry operations, because of their frequently unconventional nature, need such leaders if they are to be conducted properly. If leaders are unable to adjust to the light infantry style, tactical success may elude them. Dogmatism, inflexibility, and lack of imagination will doom light infantry forces.

The leadership at the top sets the tone of light infantry training programs and develops the tactical style that characterizes units during actual operations. This process of leadership carries down to the lower-level light infantry leaders, who conform to a broad pattern. Light infantry leaders typically are innovative, imaginative, flexible, and tough minded. They endure the same hardships as their men, so they are equally fit and self-reliant. The confidence, trust, and closeness between the leaders and those led in light infantry units normally exceed that experienced in regular infantry units. Light infantry soldiers demand a lot from their leaders, and the leaders earn respect through their performance—not by virtue of their rank.

Because of the prevalence of decentralized operations by light infantry, the quality of leadership by NCOs and junior officers is critical to success. These leaders typically are given wider latitude and more responsibility than their counterparts in regular infantry units. They mature rapidly and practice a high degree of initiative.

Furthermore, light infantry leaders demonstrate advanced technical expertise in all infantry and some special skills. For example, they must be able to fire and maintain every unit weapon, navigate to Ranger standards, read trail signs, adjust artillery fire, emplace demolitions, coordinate resupply by air, camouflage themselves and their units, and direct their units tactically. The best light infantry leaders, like First Lieutenant Logan Weston of Galahad, are "infantry scientists."[5]

Light infantry leaders are generally of higher quality than their counterparts in conventional infantry and characteristically employ several significant techniques. First, light infantry NCOs and officers lead from the front. Consequently, light forces often suffer a higher ratio of casualties in the upper ranks than other units. Second, light infantry leaders also exhibit an obvious and sincere concern about the welfare of their men. While such concern is by no means limited to the light infantry, light infantry leaders do seem to place a higher value on troop welfare than leaders in many other units. Occasionally, they go to great lengths to preserve the confidence and morale of their men. An example of this trait was Frederick's order for the delivery of a whiskey ration to his men on Difensa. Other examples are Calvert's insistence that NCOs account for every man in their squad whether living, wounded, or dead; the Chindit airgraph service; and the CCF emphasis on comradely relations and fair, equitable treatment.

The mistreatment of Galahad by Stilwell and his staff, on the other hand, illustrates the terrible consequences of paying insufficient attention to troop welfare. One can also go too far in the other direction, as the U.S. Army did in Vietnam, delivering cold beer and ice cream to units on active operations in the field. Light infantry leaders are sensitive to the needs of their men, but they do not pamper them. Thus, while they insist on immediate casualty evacuation, they do not hesitate to extend a patrol or ambush several days in time if it is tactically prudent to do so.

Light infantry leaders also typically make a point to keep their men well informed on tactical situations and the part that they will play in imminent operations. Such briefings increase the trust between the officers and the men, reduce feelings of uncertainty, and raise the level of commitment of the unit to the coming action. Furthermore, if command and control breaks down for some reason, soldiers understand the purpose of the operation and are able to carry on using their own initiative. Preoperation briefings have high value for light infantry operations.

Logistics

The central theme of the logistical philosophy of light infantry is simple: light infantry forces recognize the importance of logistics, but they refuse to be tied—either physically or mentally—to lines of communication. For light infantry, logistical planning influences, but it does not control, operational planning. Light infantrymen figure that in a pinch, they can always improvise; if necessary, they can do without.

To support themselves, light infantry forces often make maximum use of local resources. They employ the local population for certain kinds of labor, they eat the foods that nature (or natives) provide, they use natural materials for camouflage and protection, and they use the enemy's food, weapons, and ammunition against him. As masters of the environment, light infantrymen know how to exploit nature for their own sustainment.

Light infantry also improvises to simplify or solve its logistical requirements. It is always looking for lighter and better equipment or for natural substitutes. The use of elephants by the Chindits as pack animals and to clear landing zones is an example. The manner in which the 82d Airborne Division used civilian vehicles for transport in Grenada is a more modern example.

When light infantrymen transport items on their persons, specific loads are not prescribed. Individual loads vary widely based on the factors of METT-T. The Chindits had to carry about seventy pounds per man in Burma, but the SAS in Borneo insisted that their rucksacks weigh no more than fifty pounds. In the 1982 Falkland Islands War, the situation demanded that soldiers carry an average of more than 100 pounds per man. Within a given theater, for a specific campaign, however, loads can be standardized. Several principles govern the establishment of such a standard soldier's load.

Light infantrymen must be trained to carry only what is essential; NCOs and junior officers must ruthlessly restrict what soldiers put in their rucksacks. Experience will help train the men, but leaders must constantly check and correct the loads. Also, every effort must be made to lighten the soldier's load through technology and ingenuity (such as lighter rations, weapons and ammunition, and radios). Leaders at high levels must make a point of responding to the ideas of their subordinates on this matter. In addition, when local situations change, SOPs need to change. Above all, light infantrymen must not be so loaded down that they are continuously exhausted, inattentive, and unready.

Light infantry logistics have been enhanced to a great extent by the development of the helicopter. The flexibility of the helicopter, its ability to fly almost anywhere carrying heavy loads, permits the light infantry to operate independently at distances well removed from supply bases. In certain scenarios, the low daily requirement for supplies of light infantry will enable it to be resupplied completely by air. But in the absence of helicopters, current light infantry forces may find themselves relying, once again, on pack animals.

Intelligence

Each case study in this report demonstrates that accurate, timely intelligence is vital to the success of light infantry operations. To be effective, light infantry forces must know what the enemy is about, while keeping the enemy in the dark about their own intentions. While light infantry cannot afford to be surprised, it must constantly attempt to achieve surprise. To further this goal, light infantry forces obtain intelligence from sources as high as the national level (for example, the case of U.S. light forces in Grenada and the Israeli forces at Entebbe) or at the local-citizen (Malaya) or aborigine level (Burma, Borneo).

Light infantry units obtain tactical intelligence from a variety of sources. Often, the majority of their intelligence comes through comprehensive patrolling. Each one of the light forces described in the previous chapters spent a great deal of time patrolling to meet their own needs for intelligence. Light infantry also taps into existing intelligence networks, rather than attempting to duplicate them. Therefore, smooth coordination with civil and police intelligence is absolutely essential in low-intensity conflicts, particularly in counterinsurgencies. If this meshing does not take place properly, military operations may well be futile.

Other local sources of information are also employed by the light infantry: light infantry leaders use local guides, when necessary, in unfamiliar terrain; they commission border crossers to collect information on enemy dispositions; they may even form irregular units for the specific purposes of early warning and the collection of information. Sensitivity to intelligence remains an imperative for light infantry operations.

Technology

Historically, advantages in technology have not normally been decisive in light infantry operations. However, in several instances, technology has compensated for weaknesses or permitted light infantry to perform on a scale or level not previously possible. As able and self-sufficient as the Chindits were, they could not have operated at the depths that they did, nor for as long as they did, without the technology embodied in the radio and the airplanes of the No. 1 Air Commando. Similarly, if Walker had not had his helicopters in Borneo, he would have needed at least twice as many troops as he was given. Lack of technology (few airplanes, scarce transport, and few radios) certainly contributed to the failure of the CCF to drive the UN Command out of Korea, but the enormous technological superiority of the UN Command did not permit it to achieve victory over its backward foe. Stalemate resulted instead.

Light infantry leaders are cautious about how they employ available technology. Just because a technology exists does not mean that it should be used, particularly in low-intensity conflict. For example, indirect fires often accomplish little in the jungle. Also, sophisticated equipment is of little use, unless it can be man packed and handled roughly. The guiding principle for the employment of advanced technology in light infantry combat is that the technology must conform to the light infantry style and not the reverse. Light infantry leaders know this well. Overreliance on technology may rob the light infantry of its strengths. Such a practice erodes the necessary light infantry attitude of self-reliance; furthermore, it alienates the soldier from his tactical environment by creating a distracting and surreal atmosphere. The FSSF's analysis of its operations in Italy specifically cautioned against overreliance on technology. A final comment on the dangers of overreliance on technology is offered by General Nguyen Xuan Hoang. Describing the Battle of Ia Drang in the central highlands of Vietnam, Hoang stated:

> The 1st Cavalry came out to fight us with one day's food, a week's ammunition. They sent their clothes back to Saigon to be washed. They depended on water in cans, brought in by helicopter. . . . We tried to turn these advantages against you, to make you so dependent on them that you would never develop the ability to meet us on your own terms—on foot, lightly armed, in the jungle.[6]

In short, technology should be tailored to the needs of the light infantry. It should lighten the soldier's load, enhance his mobility, reduce his logistic problems, compensate for his weaknesses, nullify the enemy's advantages, but *never* alter the basic nature of the light infantry's attitude of self-reliance.

Low-Intensity Conflict

Clearly, light infantry operations in low-intensity conflicts are inherently more demanding and difficult than those in mid- to high-intensity wars. While politically derived restrictions on the use of force hinder all military operations to a significant degree, in low-intensity conflicts, they are especially constraining. In addition, light forces have the disadvantage that their enemy is hard to identify, while they, themselves, are always identifiable. In low-intensity conflicts, a higher degree of cooperation between civil, military, and police organizations is necessary for success, yet is more difficult to obtain.

Low-intensity conflicts require more patience—patience to wait for intelligence to mature, patience to accept frequent failure, patience to understand the peculiarities of the native population and government, and patience to take a long-term view on bringing a conflict to an end. The mental stress in low-intensity conflicts is also greater than that experienced in mid- to high-intensity conflicts: the enemy could be anywhere; there are no secure areas, no front and rear. This realization creates and maintains a high degree of tension.

To engage in low-intensity conflict, light infantrymen need a number of special skills and talents. Foremost among these is a sensitivity to the needs and values of the local population. "Winning hearts and minds" is more than a cliché, it embodies the essence of what the light infantryman's attitude toward the local population must be. Developing such an attitude among the rank and file of a light force must rank high on any list of priorities.

Light infantrymen also need to possess a healthy range of language skills to prepare them for low-intensity conflicts. Soldiers in such conflicts typically have much more contact with civilians than soldiers do in other forms of war. The benefits of speaking the local language in the low-intensity environment for the purposes of gaining intelligence, winning confidence, and obtaining support are self-evident.

In low-intensity conflicts, light infantrymen may also be required to apprehend suspects, conduct searches, seize property, identify contraband, man roadblocks, and support police and security forces. These tasks require a wide range of knowledge (for example, of local regulations) and skills (such as search techniques), many of which are not routinely provided for in light infantry training. Light forces may even be involved in training local security forces.

Preparing light forces for action in low-intensity conflicts often takes more time and is more difficult than developing forces for mid-intensity war, because operations in low-intensity conflicts require a wider range of skills, require more flexibility, and generate more stress in soldiers. Although such conflicts produce fewer casualties than other forms of war, the demands placed on light forces are inherently greater. Nevertheless, the light infantry ethic and low-intensity conflict are quite compatible. Because of the versatility of light forces, they adapt well to such operations.

Problems

Light infantry forces are not general-purpose forces. They have only limited use in a mid- to high-intensity war. Light forces are vulnerable to enemy artillery and aircraft fires. Moreover, they are unsuitable for sustained defensive operations because they lack the logistical infrastructure necessary to survive such operations. In addition, they lack the firepower and sustainability to attack fortified positions, except when they have perfect surprise. Although light infantry has excellent tactical mobility in close terrain, in open terrain, light forces can be outmaneuvered and outgunned with ease. Light forces always require significant support in prolonged campaigns, in open terrain, in the defense, and whenever they are pitted against heavy forces.

232

Consequently, when light forces are employed in mid- to high-intensity wars that occur in areas of varied, but primarily open, terrain, opportunities for their prudent use will be quite limited. On the other hand, in areas of operations like Korea or northern Italy, commanders will find many more chances to use light forces. Even then, however, analysis of combat operations shows that light infantry units appear to be most useful when employed at brigade level and lower, although there are exceptions to this generalization (for example, the use of airborne divisions in a coup de main strategy).

Unfortunately, there is no guarantee that higher commanders will know how to use light forces properly. Too often, higher commanders have misused light forces, even to the point of disaster. Stilwell, obviously, did not understand the capabilities and limitations of the Chindits. General Clark seemed to view the FSSF as just another brigade. Clark's corps commanders also repeatedly used the FSSF in unwise daylight assaults. SAS leaders continually resisted commanders who wished to use them in standard infantry roles.

As a result, whenever light infantry forces arrive in theater, a grave probability exists that they will be misused. Light forces are commonly misused as spearheads in ground attacks: both the FSSF and Ranger battalions served in this role during the Italian campaign. They were also misemployed in static defensive positions (as were airborne battalions and regiments). The Chindits suffered terrible casualties in daylight assaults when they were used without strong artillery support against the entrenched Japanese. The fact is higher commanders have been loath to permit high-quality forces to languish in rest areas waiting for a suitable mission to arise. So, they often order their light forces into the line or use them for other dubious ventures. If light forces are available, they will be used—rightly or wrongly. Although the misuse of light forces is deplorable, allowing the units to remain idle is expensive. In come cases, commanders have had no choice but to use light forces as conventional infantry, such as when General Bradley employed the XVIII Airborne Corps to plug the gap created by the Germans in the Ardennes counteroffensive in December 1944.

Thus, a balance must be struck in each theater of war in the use of conventional and light infantry forces. If there are too many light forces, their misuse is inevitable. If too few light forces are available, conventional forces will have to be employed in situations for which they are ill trained. During the last fifty years, whenever light and specialty forces have proliferated beyond necessary levels, it has contributed to their misuse in conventional roles.

The ultimate results of the misuse of light forces are high casualty rates and the loss of the light infantry arm. Because light forces are composed of highly trained soldiers and above-average leaders, it is difficult to obtain suitable replacements for them in combat. Unless a light infantry training base and replacement pool exist, this replacement problem cannot be solved. Eventually, disbandment may be the only alternative. The destruction of the Rangers at Cisterna is an example—although an extreme one—of the costs of misusing a light force.

The longer light forces remain in theater, the heavier they tend to become. During its history, the FSSF acquired its own airborne artillery battalion and a Ranger cannon battery. The Rangers in North Africa and Italy assimilated a 4.2-inch chemical mortar battalion. In just a few months of combat, the 10th Light (Mountain) Division acquired a collection of American and German transport, heavy machine guns, and artillery. These kinds of organizational changes are not necessarily dangerous or undesirable. They may, in fact, simply reflect bona fide requirements for extra combat power and combat support as dictated by tactical situations.

Nonetheless, light infantry forces can remain "light" in employment even though they retain relatively heavy organizations. The German mountain divisions of World War II operated in accordance with light infantry principles, yet they were larger than the standard German infantry division. Similarly, British commando units today practice the light infantry ethic, yet their organization is 100 percent mobile in armored and wheeled vehicles. When deployed, they leave behind what they do not need.

Although most light infantry forces are organized light, it is not organization that determines their light nature. It is, instead, their characteristics and methods of operation. Thus, the historical tendency for light forces to become heavier should not automatically be criticized. The danger occurs only when the tendency is uncontrolled. Then light forces can become unwieldy and inflexible, unsuited for the purposes for which they were created.

Light infantry forces are unique. Although they share many of the same skills as regular infantry, they are especially distinguished by their attitude of self-reliance, their mastery of the environment, their versatility, and their high esprit. These characteristics produce a special tactical approach to the battlefield. Offensively oriented, flexible, adaptable, and innovative, light infantry capitalizes on stealth, surprise, speed, and shock. Not psychologically tied to a supply line or to the availability of combat support, light infantry operates at night, hitting the enemy hard when and where he does not expect it. Light infantry relies on its own resources and its own organic weapons to destroy the enemy at close range. Light infantry believes that the light infantryman is the decisive weapon.

However, light infantrymen are not supermen. They get tired, become sick, and lose their effectiveness like other soldiers. Improperly used, they will die at alarming rates. On the other hand, employed by enlightened commanders and imbued with the light infantry ethic, they can be a formidable arm in time of war. (For a distilled analysis of conventional and light infantry forces, see table 7.)

Table 7. Historical Norms for Conventional and Light Infantry Forces

This table provides the distinctive differences between conventional and light infantry. The comments are intentionally terse and brief. One could probably dispute each point by reference to some historical infantry operation or complain that the distinctions drawn are too sharp and somewhat artificial; nevertheless, taken as a whole, the table conveys a general impression of those features of light infantry that distinguish it from conventional infantry. In actual operations, the differences may be blurred, and an actual overlapping of qualities may exist. In citing the distinctions between the two types of infantry, no disparagement of conventional infantry is intended. Rather, both light and conventional infantry have their necessary places on the battlefield.

Conventional Infantry	Light Infantry
Training	
Low attrition rate during training	High attrition rate during training
Mild physical demands	Extremes of physical fitness required
Weapons familiarity	Masters of weapons
Operations	
General purpose force	Utility is limited to specific conditions
Equally suited to the offense and defense	Strong offensive orientation
Operated in any terrain	Best suited for close terrain
Limited capability for unconventional operations	Adapts well to unconventional operations
Views difficult terrain as an obstacle	Dominates the terrain and uses it to its advantage
Uncomfortable in extreme climates	In harmony with the environment; adapts to nature
Operates in large formations	Most often operates at battalion level and lower
Habitually conducts daytime operations	Operates most frequently at night
Possesses built-in protection against small-arms and indirect fires (mechanized infantry only)	Achieves protection through camouflage, maneuver, and by digging in
Avoids contact with irregular forces	Frequently operates with irregular forces and special operations forces
Usually avoids contact with civilians	Makes frequent contacts with civilians for intelligence and support
Can reduce fortified positions	Ill-suited for attacks against fortified positions
Produces its own intelligence or obtains it from higher headquarters	Taps into all existing intelligence networks
Adapts to low-intensity conflict with difficulty	Naturally suited for low-intensity conflict
Operates as part of a large combined arms formation	Usually operates in a pure infantry environment
Tactics	
Employs conventional tactics per field manuals	Employs unusual tactics, usually adapted specifically to the environment
Seeks 3:1 advantage in mass and firepower in the attack	Often fights on equal terms, sometimes outnumbered
Mass is the primary tactical principle	Surprise is the primary tactical principle
Achieves shock through mass	Achieves shock through surprise, speed, and violence
Relies on artillery preparations	Frequently employs no artillery preparations
Follows the path of least resistance	Chooses the path of least expectation
Uses roads and trails	Avoids roads and trails
Engages the enemy at maximum range	Engages the enemy at close range
Defends on the forward slope	Defends from the reverse slope
Normally emphasizes firepower over maneuver	Emphasizes maneuver over firepower
Excellent mobility in open and mixed terrain	Can be outnumbered in open terrain
Low mobility in close terrain	Excellent mobility in close terrain

Frequently conducts frontal assaults	Infiltrates in order to attack the enemy's flank and rear
Employs camouflage to enhance survivability	Expert camouflage is a matter of life and death
Patrols to maintain contact	Patrols relentlessly in all situations
Tactics conform to a general pattern	Tactically unpredictable in form, time, and space
Weapons and equipment oriented	People and terrain oriented
Adjusts tactics to available technology	Adjusts available technology to tactics

Combat Support

Depends heavily on strong combat support	Relies primarily on its own organic weapons
Basic organization includes a balance of arms and services	Basic organization includes few combat-support elements. Acquires such support on a temporary basis.

Logistics

Physically and psychologically dependent on fixed lines of communication	Self-reliant; operates independently of fixed lines of communication
Basic organization includes a robust combat service support tail	Basic organization includes few combat service support elements
Depends on formal logistics structure	Improvises to meet needs; uses local and enemy resources
Can sustain itself in attrition warfare	Lacks sustainability for attrition warfare
Comfort conscious	Practices self denial
Has heavy daily logistics requirements	Routinely practices austerity
May not closely regulate the soldier's loads	Establishes strict SOPs on soldiers' loads
Cannot operate far from lines of communications and supply bases	Austerity and improvisation permit operations far removed from supply bases
Resupplied by air only with difficulty	Often resupplied by air because of low daily requirements

Leadership

Centralized tactical direction	Decentralized responsibility; wide latitude granted NCO's and junior officers
Tactical initiative employed within the limits of the overall operation	Practices innovation, imagination, and initiative to a high degree
Adequate technical expertise	Infantry scientists
Values troop welfare	High sensitivity to troop welfare
Infrequent troop briefings	Troops kept constantly informed

NOTES

Chapter 5

1. Edward N. Luttwak, et al., *Historical Analysis and Projection for Army 2000*, Pt. 1, Paper no. 16, *Notes on the Israeli 35th (Paratroop) Brigade and Derived Reserve Brigades, with Additional Notes on the 'Air-Landed Force' and the Golani Brigade* (Chevy Chase, MD: Edward N. Luttwak, 1 March 1983).

2. The light divisions in the 1985 "army of excellence" force structure do *not* possess a "break-in" capability. Thus, they cannot be deployed unless a secure lodgment already exists or unless a lodgment is first secured by other forces.

3. The Japanese employed formidable reverse-slope defense works in the Battle of Okinawa in 1945 in the Pacific war. U.S. divisions breached these defenses only after suffering very high casualties during days of dogged, close-in fighting. This campaign is described well in Roy E. Appleman, et al., *Okinawa: The Last Battle*, United States Army in World War II (1948; reprint, Washington, DC: Historical Division, Department of the Army, 1977). A good recent study on reverse-slope defenses is Lieutenant Colonel Archibald Galloway's "Light Infantry in the Defense: Exploiting the Reverse Slope from Wellington to the Falklands and Beyond," unpublished monograph for the School of Advanced Military Studies, U.S. Army Command and General Staff College, Fort Leavenworth, KS, 2 December 1985.

4. For example, the Israeli airborne brigade retains armored personnel carriers in its battalions, the British 3 Commando Brigade employs vehicles from the Spartan family of armored cars, as does the British 5 Infantry Brigade (Worldwide Tasks) (Airborne). U.S. separate light brigades also include an armored cavalry organization.

5. Charlton Ogburn, *The Marauders* (New York: Harper & Brothers, 1956), 39—41.

6. General Nguyen Xuan Hoang, "A Veteran Returns," *Army Times*, 6 May 1985.

237

BIBLIOGRAPHY

Chapter 5

Appleman, Roy E., et al. *Okinawa: The Last Battle*. United States Army in World War II. 1948. Reprint. Washington, DC: Historical Division, Department of the Army, 1977.

Galloway, Archibald, Lieutenant Colonel. "Light Infantry in the Defense: Exploiting the Reverse Slope from Wellington to the Falklands and Beyond." Unpublished monograph for the School of Advanced Military Studies, U.S. Army Command and General Staff College, Fort Leavenworth, KS, 2 December 1985.

Hoang, Nguyen Xuan, General. "A Veteran Returns." *Army Times*, 6 May 1985.

Luttwak, Edward N., et al. *Historical Analysis and Projection for Army 2000*. Pt. 1. Paper no. 16. *Notes on the Israeli 35th (Paratroop) Brigade and Derived Reserve Brigades, with Additional Notes on the 'Air-Landed Force' and the Golani Brigade*. Chevy Chase, MD: Edward N. Luttwak, 1 March 1983.

Ogburn, Charlton. *The Marauders*. New York: Harper & Brothers, 1956.

✿ U.S. GOVERNMENT PRINTING OFFICE: 1987 554-001/62108

ABOUT THE AUTHOR

Major Scott R. McMichael graduated with honors from Davidson College in 1972 (BA, history) and was commissioned through the ROTC into the Field Artillery. In 1978, he received a master's degree in international relations from the University of Chicago. A graduate of airborne and Ranger schools, he has served in artillery assignments, including duties as a battery commander and battalion operations and training officer (S3). From 1983 to 1986, he was assigned to the Combat Studies Institute, USACGSC, where he researched and wrote this research survey. He is now studying Russian at the Defense Language Institute, Monterey, California, preparatory to going to the U.S. Army Russian Institute in Garmisch, Federal Republic of Germany.

Scott R. McMichael

COMBAT STUDIES INSTITUTE

Missions

The Combat Studies Institute was established on 18 June 1979 as a department-level activity within the U.S. Army Command and General Staff College, Fort Leavenworth, Kansas. CSI has the following missions:

1. Conduct research on historical topics pertinent to the doctrinal concerns of the Army and publish the results in a variety of formats for the Active Army and Reserve Components.

2. Prepare and present instruction in military history at USACGSC and assist other USACGSC departments in integrating military history into their instruction.

3. Serve as the U.S. Army Training and Doctrine Command's executive agent for the development and coordination of an integrated, progressive program of military history instruction in the TRADOC service school system.

www.ingramcontent.com/pod-product-compliance
Lightning Source LLC
Chambersburg PA
CBHW062038090426
42740CB00016B/2949